ANNALS *of* THE NEW YORK ACADEMY OF SCIENCES

T0179975

VOLUME
1287

ISBN-10: 1-57331-930-0

ISSUE

Annals Meeting Reports

The conference "Application of Combined 'omics Platforms to Accelerate Biomedical Discovery in Diabesity" was presented by Hot Topics in Life Sciences and the Sackler Institute for Nutrition Science at the New York Academy of Sciences.

The conference "Prioritizing Health Disparities in Medical Education to Improve Care" was presented by the Josiah Macy Jr. Foundation, the Associated Medical Schools of New York, the New York University School of Medicine, and the New York Academy of Sciences.

The conference "The Paradox of Overnutrition in Aging and Cognition, Precursors of Aging: Facilitating Intervention Strategies" was presented by the Sackler Institute for Nutrition Science at the New York Academy of Sciences.

The conference "Vitamin D: Beyond Bone," presented by the Sackler Institute for Nutrition Science at the New York Academy of Sciences, was sponsored by an unrestricted educational grant from Abbott Nutrition Health Institute.

TABLE OF CONTENTS

EDITOR-IN-CHIEF
Douglas Braaten

ASSOCIATE EDITORS
David Alvaro
Azra Jaferi

PROJECT MANAGER
Steven E. Bohall

DESIGN
Ash Ayman Shairzay

The New York Academy of Sciences
7 World Trade Center
250 Greenwich Street, 40th Floor
New York, NY 10007-2157
annals@nyas.org
www.nyas.org/annals

Annals of the New York Academy of Sciences (ISSN: 0077-8923 [print]; ISSN: 1749-6632 [online]) is published 30 times a year on behalf of the New York Academy of Sciences by Wiley Subscription Services, Inc., a Wiley Company, 111 River Street, Hoboken, NJ 07030-5774.

Mailing: *Annals of the New York Academy of Sciences* is mailed standard rate.

Postmaster: Send all address changes to ANNALS OF THE NEW YORK ACADEMY OF SCIENCES, Journal Customer Services, John Wiley & Sons Inc., 350 Main Street, Malden, MA 02148-5020.

Disclaimer: The publisher, the New York Academy of Sciences, and the editors cannot be held responsible for errors or any consequences arising from the use of information contained in this publication; the views and opinions expressed do not necessarily reflect those of the publisher, the New York Academy of Sciences, and editors, neither does the publication of advertisements constitute any endorsement by the publisher, the New York Academy of Sciences and editors of the products advertised.

Publisher: *Annals of the New York Academy of Sciences* is published by Wiley Periodicals, Inc., Commerce Place, 350 Main Street, Malden, MA 02148; Telephone: 781 388 8200; Fax: 781 388 8210.

Journal Customer Services: For ordering information, claims, and any inquiry concerning your subscription, please go to www.wileycustomerhelp.com/ask or contact your nearest office. *Americas:* Email: cs-journals@wiley.com; Tel:+1 781 388 8598 or 1 800 835 6770 (Toll free in the USA & Canada). *Europe, Middle East, Asia:* Email: cs-journals@wiley. com; Tel: +44 (0) 1865 778315. *Asia Pacific:* Email: cs-journals@wiley.com; Tel: +65 6511 8000. *Japan:* For Japanese speaking support, Email: cs-japan@wiley.com; Tel: +65 6511 8010 or Tel (toll-free): 005 316 50 480. Visit our Online Customer Get-Help available in 6 languages at www.wileycustomerhelp.com.

Information for Subscribers: *Annals of the New York Academy of Sciences* is published in 30 volumes per year. Subscription prices for 2013 are: Print & Online: US$6,053 (US), US$6,589 (Rest of World), €4,269 (Europe), £3,364 (UK). Prices are exclusive of tax. Australian GST, Canadian GST, and European VAT will be applied at the appropriate rates. For more information on current tax rates, please go to www.wileyonlinelibrary.com/tax-vat. The price includes online access to the current and all online back files to January 1, 2009, where available. For other pricing options, including access information and terms and conditions, please visit www.wileyonlinelibrary.com/access.

Delivery Terms and Legal Title: Where the subscription price includes print volumes and delivery is to the recipient's address, delivery terms are Delivered at Place (DAP); the recipient is responsible for paying any import duty or taxes. Title to all volumes transfers FOB our shipping point, freight prepaid. We will endeavour to fulfill claims for missing or damaged copies within six months of publication, within our reasonable discretion and subject to availability.

Back issues: Recent single volumes are available to institutions at the current single volume price from cs-journals@wiley.com. Earlier volumes may be obtained from Periodicals Service Company, 11 Main Street, Germantown, NY 12526, USA. Tel: +1 518 537 4700, Fax: +1 518 537 5899, Email: psc@periodicals.com. For submission instructions, subscription, and all other information visit: www.wileyonlinelibrary.com/journal/nyas.

Production Editors: Kelly McSweeney and Allie Struzik (email: nyas@wiley.com).

Commercial Reprints: Dan Nicholas (email: dnicholas@wiley.com).

Membership information: Members may order copies of *Annals* volumes directly from the Academy by visiting www. nyas.org/annals, emailing customerservice@nyas.org, faxing +1 212 298 3650, or calling 1 800 843 6927 (toll free in the USA), or +1 212 298 8640. For more information on becoming a member of the New York Academy of Sciences, please visit www.nyas.org/membership. Claims and inquiries on member orders should be directed to the Academy at email: membership@nyas.org or Tel: 1 800 843 6927 (toll free in the USA) or +1 212 298 8640.

Printed in the USA by The Sheridan Group.

View *Annals* online at www.wileyonlinelibrary.com/journal/nyas.

Abstracting and Indexing Services: *Annals of the New York Academy of Sciences* is indexed by MEDLINE, Science Citation Index, and SCOPUS. For a complete list of A&I services, please visit the journal homepage at www. wileyonlinelibrary.com/journal/nyas.

Access to *Annals* is available free online within institutions in the developing world through the AGORA initiative with the FAO, the HINARI initiative with the WHO, and the OARE initiative with UNEP. For information, visit www. aginternetwork.org, www.healthinternetwork.org, www.oarescience.org.

Annals of the New York Academy of Sciences accepts articles for Open Access publication. Please visit http://olabout.wiley.com/WileyCDA/Section/id-406241.html for further information about OnlineOpen.

Wiley's Corporate Citizenship initiative seeks to address the environmental, social, economic, and ethical challenges faced in our business and which are important to our diverse stakeholder groups. Since launching the initiative, we have focused on sharing our content with those in need, enhancing community philanthropy, reducing our carbon impact, creating global guidelines and best practices for paper use, establishing a vendor code of ethics, and engaging our colleagues and other stakeholders in our efforts. Follow our progress at www.wiley.com/go/citizenship.

ANNALS OF THE NEW YORK ACADEMY OF SCIENCES
Issue: Annals *Meeting Reports*

Application of combined omics platforms to accelerate biomedical discovery in diabesity

Irwin J. Kurland,[1] Domenico Accili,[2] Charles Burant,[3] Steven M. Fischer,[4] Barbara B. Kahn,[5,6] Christopher B. Newgard,[7] Suma Ramagiri,[8] Gabriele V. Ronnett,[9] John A. Ryals,[10] Mark Sanders,[11] Joe Shambaugh,[12] John Shockcor,[13] and Steven S. Gross[14]

[1]Department of Medicine, Stable Isotope and Metabolomics Core Facility, Albert Einstein College of Medicine Diabetes Center, Bronx, New York. [2]Diabetes and Endocrinology Research Center, Columbia University, New York, New York. [3]Department of Internal Medicine, University of Michigan Medical School, Ann Arbor, Michigan. [4]Agilent Technologies, Santa Clara, California. [5]Beth Israel Deaconess Medical Center and Harvard Medical School, Boston, Massachusetts. [6]Daegu Gyeongbuk Institute of Science and Technology, Daegu, Korea. [7]Sarah W. Stedman Nutrition and Metabolism Center, Duke University Medical Center, Chapel Hill, North Carolina. [8]AB SCIEX, Concord, Ontario, Canada. [9]Department of Neuroscience, Johns Hopkins University School of Medicine, Baltimore, Maryland. [10]Metabolon, Inc., Durham, North Carolina. [11]Thermo Fisher Scientific, Somerset, New Jersey. [12]Genedata Inc., San Francisco, California. [13]Waters Corporation, Milford, Massachusetts. [14]Department of Pharmacology, Weill Cornell Medical College, New York, New York

Address for correspondence: Irwin J. Kurland M.D. Ph.D., Department of Medicine, Stable Isotope and Metabolomics Core Facility, Michael F. Price Center, 1301 Morris Park Avenue, Room 374, Bronx, NY 10461

Diabesity has become a popular term to describe the specific form of diabetes that develops late in life and is associated with obesity. While there is a correlation between diabetes and obesity, the association is not universally predictive. Defining the metabolic characteristics of obesity that lead to diabetes, and how obese individuals who develop diabetes different from those who do not, are important goals. The use of large-scale omics analyses (e.g., metabolomic, proteomic, transcriptomic, and lipidomic) of diabetes and obesity may help to identify new targets to treat these conditions. This report discusses how various types of omics data can be integrated to shed light on the changes in metabolism that occur in obesity and diabetes.

Keywords: omics; diabesity; diabetes; obesity; metabolomics; proteomics; lipidomics; metabolism, metabolic profiling

Introduction

Diabetes is an increasing concern not only for Western countries, where diet and lifestyle promote expanding waistlines and insulin resistance, but also for developing countries in which the effects of changing diet on the health of their populations are already visible. In the U.S., diabetes affects approximately 11% of the population over age 20, and there are an additional 79 million adults with pre-diabetes, a condition that often precedes diabetes in which glucose levels are higher than normal.[1]

Diabetics suffer an impairment of the body's ability to switch between glucose and fat as energy sources. Normally, when a person has not eaten recently (a fasting state), the muscles preferentially oxidize fat over glucose to ensure a supply of glucose for the brain. After a person eats, however, there is excess glucose in the system, and the muscles switch their primary energy source and begin oxidizing glucose and storing fats. Even early in the evolution of diabetes (i.e., in the pre-diabetic state referred to as metabolic syndrome), individuals are unable to make this fuel switch, a physiological maladaptation termed *metabolic inflexibility*.[2,3] Muscles that use too much glucose in the fasted state contribute to fasting hyperlipidemia, and muscles that continue to oxidize fats in the fed state, instead of switching to glucose utilization, contribute to post-prandial hyperglycemia. Muscle metabolic inflexibility, along with the failure of insulin to suppress fat breakdown and post-prandial hepatic glucose production in the pre-diabetic and diabetic states (insulin resistance), results in high blood lipid and glucose levels.

doi: 10.1111/nyas.12116

As diabetes and its associated comorbidities—such as cardiovascular disease, kidney disease, and neurological disorders—rise in epidemic proportions, it is now more important than ever to develop new tools to understand the complex metabolic mechanisms and pathways involved in this disease and to find new therapeutic targets. In April 2012, leaders in this field met at the New York Academy of Sciences to discuss how various types of omics data (metabolomic, proteomic, transcriptomic, and lipidomic) can be integrated to reveal a more complete picture of these mechanisms.

The primary focus of the conference "Application of Combined 'omics Platforms to Accelerate Biomedical Discovery in Diabesity" was obesity-induced diabetes—*diabesity*, which covers a constellation of signs, including obesity, insulin resistance, metabolic syndrome, and diabetes.[4,5] Not all obese people have diabetes and not all people with diabetes are obese, but there is definitely a connection between the two conditions. One of the main questions throughout the conference was how to use omics data to create a phenotypic profile of disease state progression in order to understand why some individuals develop diabetes and its associated complications, while others do not.

New tools and frameworks for gathering and visualizing omics data

As an alternative to shotgun accumulation of large omic data sets, phenotypic data gathering can be done in a step-wise progression for hypothesis driven research. Irwin Kurland (Albert Einstein College of Medicine) presented a tiered framework in which commonly used measures of metabolism (e.g., phenotyping tests such as calorimetry and body composition analysis) and a novel deuterated glucose tolerance test (termed the hepatic recycling deuterated glucose tolerance test, or HR-dGTT) that assesses peripheral versus hepatic glucose disposal,[6,7] are performed first to help determine which specific omics experiments to do next in animal models (Fig. 1). The results of each of these tests can inform subsequent experiments to generate a hypothesis-driven, multi-omic investigative framework.

If the measurements of fuel utilization by indirect calorimetry (Fig.1, panel I) indicate a change in carbohydrate or fat utilization, for example, plasma and muscle metabolomic and lipidomic profiling

may be indicated. Or, if measurements of body composition reveals changes in body fat (Fig. 1, panel II), one could follow up by measuring lipogenesis using deuterated water[8] or lipolysis using $[2-^{13}C]$-glycerol.[8–10] Changes in lipogenesis and/or lipolysis then provide enough evidence to follow up with lipidomic analyses, such as acyl carnitine or acyl CoA profiling,[8] to monitor which lipids are being produced and/or broken down. The hepatic recycling glucose (deuterated) tolerance test (HR-dGTT, Fig. 1, panel III) assesses peripheral glucose disposal, as well as the recycling of glucose through the liver (a function of hepatic glucose uptake), based on plasma measurements that assess the decay in relative enrichment of administered $[2-^2H_1]$-glucose, versus $[6,6-^2H_2]$-glucose. Notably, while both $[2-^2H_1]$-glucose and $[6,6-^2H_2]$-glucose are taken up by the liver (via a process catalyzed by glucokinase); only $[6,6-^2H_2]$-glucose exits the liver unchanged after traveling through the glucose/glucose-6-P futile cycle (via a process catalyzed by glucose-6-phosphatase). However, there is substantial loss of $[2-^2H_1]$-glucose before exiting by exchange of the deuterium at the 2-position with water protons during the rapid equilibration of glucose-6-P with fructose-6-P, which does not affect hydrogens at carbon 6.[6] Peripheral glucose disposal is estimated from the $[6,6-^2H_2]$-glucose area under the curve (AUC) during the HR-dGTT, and insulin AUC can also be obtained for an estimate of whole body insulin resistance. If changes in hepatic versus peripheral glucose disposal are observed by the HR-dGTT, additional stable isotope tests can be performed to monitor hepatic glucose production (HGP), lipolysis and glucose/glycerol recycling (by assessing glycerol production and HGP from glycerol), and glucose/lactate (Cori) re-cycling (by assessing lactate production and HGP from lactate) (Fig. 1). The results of these tests can coordinate tissue-specific metabolomic and lipidomic profiling efforts. Decisions can then be made to perform related global omic profiling, such as tissue-specific acetylome determination, thus leading to integrated omic information that may underlie diabetes development.

This sequential phenotyping paradigm has been applied to several mouse models,[6–17] including a model of increased insulin sensitivity, the Pten[+/−] mouse[6] and a fatty acid amide hydrolase (FAAH)-knockout mouse, a novel model of the pre-diabetic

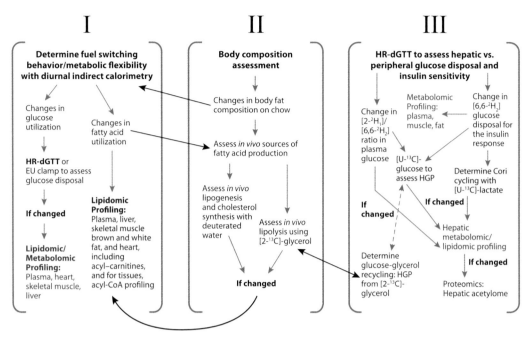

Figure 1. Framework for integrating fluxomic, metabolomic and lipidomic profiling. Our approach is to use fluxomics as a primary tool for metabolic phenotyping, and to layer additional omic information, such as metabolomics, lipidomics, and proteomics (acetylome determination), in a hypothesis-driven manner, and vice versa, to use fluxomics to elucidate the importance of other omic findings. (I) Discovery framework resulting from observing changes in fuel utilization with indirect calorimetry. (II) Discovery framework resulting from observing changes in body composition. (III) Discovery framework resulting from observing changes in flux measured via the hepatic recycling glucose (deuterated) tolerance test (HR-dGTT). The HR-dGTT yields information about peripheral and hepatic glucose disposal that can localize tissue specific metabolic/lipidomic screening. Changes in hepatic versus peripheral glucose disposal are assessed from the time course of percent differences in plasma $[2\text{-}^2H_1]$-glucose vs. $[6,6\text{-}^2H_2]$-glucose enrichments (1-ratio($[2\text{-}^2H_1]$/ $[6,6\text{-}^2H_2]$)-glucose).[6] The correlation with the hepatic global acetylome can be assessed in a hypothesis-driven framework, along with other fluxomic methodologies for assessing lipolysis (adipose), hepatic-adipose/glucose-glycerol recycling (HGP from $[2\text{-}^{13}C]$-glycerol), lipogenesis, and Cori cycling (HGP from $[U\text{-}^{13}C]$-lactate). The stable isotope tests shown are closed loop tests that are performed at the basal glucose and insulin levels (glycerol production and HGP), or dynamic tests incorporating the HR-dGTT and insulin responses. Closed loop tests do not require experimental groups to have identical, fixed values in glucose and insulin that are needed for open loop tests like the euglycemic hyperinsulinemic (EU) clamp.[7] Tissue assessments in this framework assume an animal model. HGP, hepatic glucose production; D_2O, deuterated water. Image courtesy of Irwin J. Kurland.

state,[8] which has reduced hydrolysis of endo-cannabinoids such as anadamide, and a type 2 diabetic mouse model, the MKR mouse.[10] Pten normally inhibits insulin signaling by deactivating the product of insulin-stimulated phosphatidyli-nositide 3-kinases. Because insulin signaling stimu-lates hepatic glucose uptake, one might expect that the Pten$^{+/-}$ mouse would show increased hepatic glucose uptake. However, the HR-dGTT revealed dramatically decreased hepatic glucose uptake in the Pten$^{+/-}$ mouse, which correlated with decreased basal glucokinase expression, whereas HGP was the same as in the wild-type mouse. To explain these

counterintuitive results, Kurland and collaborators hypothesized that, to ensure that enough glucose is supplied to the brain, hepatic glucose uptake is dramatically suppressed in the fasted state, leaving hepatic gluconeogenesis unaffected, so that hepatic glucose production occurs as normal. Glucokinase expression in the fasted to re-fed transition was markedly induced in Pten$^{+/-}$ mouse livers,[6] indi-cating increased insulin sensitivity, suggesting basal HGP regulation is under the control of factors be-sides insulin signaling, such as neural control.

   The second model Kurland presented was the FAAH$^{-/-}$ mouse.[8] FAAH$^{-/-}$ mice mimic several metabolic aspects of pre-diabetes, including obesity impaired fuel utilization, hyperinsulinemia, and insulin resistance in liver, skeletal muscle, and adipose tissue. The HR-dGTT indicated that the $FAAH^{-/-}$ mice had higher fasting insulin levels and higher blood glucose and insulin levels during the GTT even though glucose uptake in the periphery was the same as wild-type mice.[8] The hyperglycemia was due, in part, to a non-suppressibility of HGP, indicated by the difference between the total plasma glucose (labeled and unlabeled) level during the HR-dGTT (higher for $FAAH^{-/-}$) and the [6,6-$^2H_2$]-glucose level (unchanged versus wild type) during the HR-dGTT, as well as non-suppressibility of HGP (demonstrated with $U^{13}C$ glucose) during fasting. In accord with the stepwise phenotyping procedure described in Figure 1 (panels II and III), this led to examining the breakdown of lipids in adipose tissue (lipolysis), and HGP from glycerol as lipolysis of adipose triglycerides creates glycerol and fatty acids. Notably, administering [2-^{13}C]-glycerol to mice and monitoring its dilution can provide information on how actively adipose tissue is breaking down lipids, and [2-^{13}C]-glycerol can be followed to the production of ^{13}C-glucose to assess HGP from the hepatic triose-P pool (Fig. 1). In $FAAH^{-/-}$ mice, nonsuppressed and increased basal glycerol production and a corresponding increase in the use of glycerol for glucose production in the liver were observed. Metabolite profiling was then indicated (Fig. 1) and showed decreased triose-P metabolites in the fasted state of $FAAH^{-/-}$ mice that support the re-direction of triose-P intermediates to the increased HGP from glycerol seen. $FAAH^{-/-}$ mice also showed changes in TCA cycle metabolites that affect the malate–aspartate shuttle, one of the main conduits for transferring energy from glycolysis into the mitochondria. In particular, fasted citrate levels were decreased and fed citrate levels increased, indicating perturbations in acetyl CoA levels that were subsequently confirmed by direct measurements of acetyl carnitine and acetyl CoA measurements. This lead to the assessment of the fasted/fed hepatic acetylome following the framework shown in Figure 1.

Acetyl CoA sits on the crossroad of glucose, fatty acid, amino acid, and cholesterol metabolism, and so acetyl CoA has been proposed to be part of metabolic sensor and feedback mechanisms that regulate fuel utilization in the fasted and re-fed states (reviewed in Yang *et al.*[15]). The acetylome consists of proteins whose activities are regulated by acetylation, and this process relies on acetyl CoA as an acetyl donor. Changes in acetylation for mitochondrial malate dehydrogenase (MDH2), which was hypoacetylated in fasted $FAAH^{-/-}$ livers, and hyperacetylated in fed $FAAH^{-/-}$ livers, supports the metabolite profiling, indicating an impairment in the malate/aspartate shuttle. While dihydroxyacetone-P (DHAP) and glycerol-3-P levels were decreased in the fasted state of the $FAAH^{-/-}$ mice, they were preserved in the fed state, consistent with a compensating contribution from a decrease in fed aldolase B acetylation in $FAAH^{-/-}$ mice. These studies show how, by beginning with simple whole body measurements, such as calorimetry and measurements of body composition, one can eventually work towards understanding mechanisms at the molecular level.

Tools for evaluating omics data

The complement to gathering omic data by applying the hypothesis-driven framework of Kurland and his collaborators (Fig. 1) is to gather and use omics data in an untargeted approach to try to generate novel hypotheses and to identify new targets that inform subsequent experiments. A challenge is that untargeted omics data collections can be difficult to analyze due to the sheer size of the datasets.

Charles Burant (University of Michigan Medical School) discussed two programs developed to visualize and analyze several types of omics data. The hope is that, by using these tools, researchers can generate hypotheses about the metabolic networks that respond to particular types of intervention, which can then be tested for their therapeutic value.

The first tool, Metscape 2,[18] is a plugin for the program Cytoscape, a common platform for visualizing complex networks. Currently, Metscape 2 can incorporate gene expression and metabolomics data across different time points or different experimental conditions. Based on input data, Metscape 2 creates interaction maps that allow researchers to visualize the changes in gene expression and in metabolite levels in an attempt to link these changes to disease states. The second tool that Burant discussed was CoolMap (developed by colleagues G. Su and M. Fan), which enables researchers to visualize large, two-dimensional data and to interpret correlations between datasets. CoolMap can manage datasets of 8000 × 8000 data points and shows the Pearson's correlation coefficient in a heat map-like format.

As an example, Burant showed two CoolMap plots of various clinical parameters before and after weight loss. By visualizing the changes in these parameters, researchers can focus on the relationships that differ between the two states and can generate hypotheses that can be tested in further experiments. The usefulness of CoolMap will manifest in its ability to identify *known unknowns*, which are, according to Burant, unidentified, reproducible features in mass spectrometry data generated from untargeted high-throughput metabolomic studies (Fig. 2). To demonstrate this point, Burant used a CoolMap to show the correlation between various metabolites (fatty acids, amino acids, acetyl CoA, etc.) from a group of 25 people after the subjects had lost an average of 22.5% of their body weight. CoolMap can cluster the metabolites to reveal groups of metabolites that are highly related. By exporting a group of highly-related metabolites into Metscape 2, Burant showed that these metabolites were all part of a common pathway. Once a particular pathway is suspected of being important, researchers can hypothesize what other metabolites they should be able to see in their data and can go back to their original mass spectrometry data and identify some of their known unknowns.

While Burant and colleagues have already made Metscape 2 available to researchers (with CoolMap soon to be released), they are also constantly improving the platforms. Future versions of Metscape 2 should be able to integrate proteomic, phospho-proteomic, and acetylomic data to understand the relationship between genes, proteins, and metabolites in various states. Burant and co-workers are working to integrate CoolMap with Metscape 2 and with other omics programs to provide a suite of tools that integrate various types of directed (targeted) metabolomics, in which specific metabolites of interest are identified and quantified against stable isotope standards, as well as undirected (untargeted) metabolomics, in which researchers are not looking for specific metabolites but are instead performing an unbiased survey of which metabolites are sensitive to changes in various conditions.

Branched chain amino acids

One of the goals of omics techniques, as described by Christopher Newgard (Duke University Medical Center), is to create metabolic signatures of human diseases that can be used as prognostic factors, to monitor disease progression, guide therapeutic interventions, and for hypothesis generation that can be tested in animal models. Newgard's talk included a comparison between targeted metabolic profiling of obese individuals with pre-diabetes and insulin resistance (a body mass index (BMI) of ∼36) to that of lean individuals (a BMI of ∼22).[19] While previous studies have also looked at metabolic differences between obese and lean individuals, they have primarily focused on one or a small number of metabolites based on the particular hypothesis of each study. In contrast, Newgard's study gathered omic data to generate hypotheses using targeted metabolomics to measure over one hundred analytes.

After grouping the analytes of interest by principal component analysis (PCA), Newgard focused on one group—containing branch-chain amino acids (BCAAs-valine, leucine and isoleucine), glutamate and glutamine, 3- and 5-carbon acyl carnitines (C3-AC, C5-AC), and aromatic amino acids phenylalanine and tyrosine—that together explain most of the variance in the data. Most of these compounds are linked not just by PCA analysis but metabolically via BCAA metabolism. For example, isoleucine and leucine produce C5-AC in mitochondria from 2-methylbutyryl CoA and isovaleryl CoA, respectively. Glutamate is linked to the BCAAs via transamination in the cytoplasm. BCAAs go through a similar set of reactions during catabolism, which generate glutamate during a transamination first step in the cytoplasm; and C3-AC is generated from propionyl CoA produced from valine and isoleucine metabolism. The aromatic amino acids phenylalanine and tyrosine may compete with the BCAAs for the same transporters to enter cells.

Other studies have shown an association between BCAA levels and insulin resistance; however, the advantage of Newgard's study is that because it was done by using an unbiased metabolomic analysis, the researchers were able to show that the whole pathway related to BCAA metabolism is elevated, and that the BCAA profile was the one most strongly associated with insulin resistance: interestingly, more so than the lipid-related signatures.

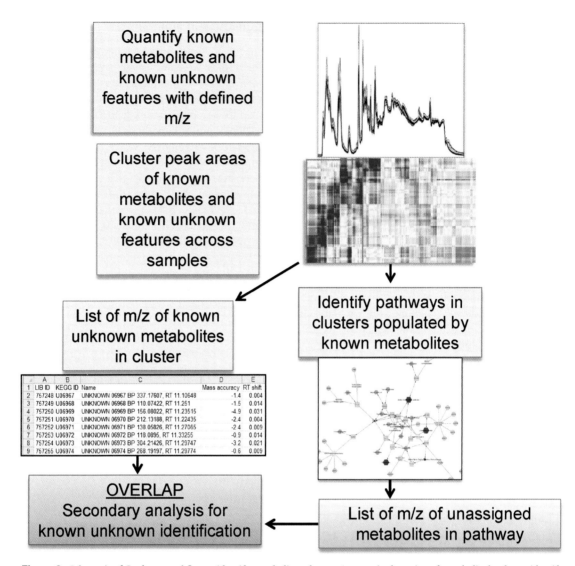

Figure 2. Schematic of Coolmap workflow to identify metabolic pathways. Automatic clustering of metabolite levels can identify metabolic pathways generated by known compounds contained in the clusters. 'Known Unknown' features with masses matching other metabolites in the identified pathway can aid in the identification of unknown metabolites. Image courtesy of Charles Burant.

Newgard demonstrated that the BCAA/insulin resistance signature replicates in other cohorts with insulin resistance, for example, the Studies of a Targeted Risk Reduction Intervention through Defined Exercise (STRRIDE) trial[20] and an Asian/Indian cohort.[21] Also, the BCAA cluster of metabolite associations in interventions to treat diabetes, such as gastric bypass surgery and weight loss, predicts who will have an improvement in insulin sensitivity more effectively than the amount of weight actually lost.[22] The decrease in BCAA levels corresponds to the efficacy of the interventions, and gastric by-pass was associated with a greater decrease in BCAA than was matched weight-loss intervention.[23] Importantly, BCAAs were shown to play a causative role in insulin resistance. Rats fed a high-fat diet supplemented with BCAAs spontaneously ate less food and weighed mildly less than rats fed a normal high-fat diet, but rats on both diets were equally insulin-resistant.[19]

Newgard proposed a mechanism for the role of BCAAs in insulin resistance (Fig. 3) that centers on the role of inter-organ flux of BCAAs. Gastric bypass patients can have low expression of

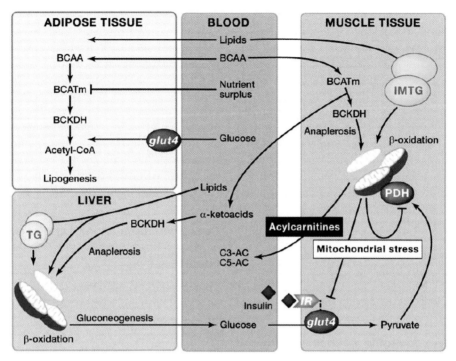

Figure 3. Schematic of a working model of potential crosstalk between lipids and branched chain amino acids (BCAA) in the development of obesity-related insulin resistance. *Anaplerosis* refers to repletion or filling up of TCA cycle intermediates via entry points other than acetyl CoA. TG, triglyceride; IMTG, intramyocellular triglyceride; IR, insulin receptor; BCATm, mitochondrial branched-chain aminotransferase; BCKDH, branched chain keto acid dehydrogenase; PDH, pyruvate dehydrogenase. Image courtesy of Christopher Newgard.

BCAA-metabolizing enzymes in adipose tissue, which increases after gastric bypass surgery and may explain the decrease in plasma BCAAs seen after gastric bypass surgery. In the setting of nutrient and caloric excess, which often occurs in a Western diet, the normal catabolism of BCAAs in adipose tissue is overwhelmed, and BCAAs exit into the bloodstream. These BCAAs find their way to muscle where they generate CoA species, such as succinyl CoA and proprionyl CoA, which enter the TCA cycle and impair the ability of mitochondria to completely oxidize fat. In the presence of these excess nutrients, the fuel-switching ability of the cell is impaired and glucose becomes almost superfluous as a fuel source, which could lead to the high blood glucose levels observed in pre-diabetes and a disturbance in metabolic fuel selection in diabetes even in the absence of impaired insulin signaling. The source of the BCAAs may also be related to the microbiome.

Barbara Kahn (Beth Israel Deaconess Medical Center and Harvard Medical School) followed up on the investigation of impaired BCAA metabolism in adipose tissue using a branched-chain aminotransferase (BCAT)-knockout mouse as a model. Knocking out BCAT impairs the ability to metabolize BCAAs and results in high serum BCAA levels. This state mimics characteristics of obesity, in which enzymes involved in BCAA metabolism are often downregulated, leading to high levels of circulating BCAAs. Replacing the adipose tissue in the BCAT-knockout mice with normal adipose tissue decreased the circulating levels of BCAAs, demonstrating that adipose tissue does indeed play a major role in regulating the levels of BCAAs.[24]

Teasing the link between diabetes and obesity with mouse models

In addition to the work on BCAAs, Newgard presented data on using mouse models in an attempt to understand the link between obesity and diabetes. Starting with two common laboratory strains of mice, C57BL/6 and BTBR, Newgard, in collaboration with Alan Attie (University of Wisconsin), introduced the *ob* gene into these mice to create two distinct strains of genetically-induced obese mice. While both strains are insulin resistant, only one progresses to diabetes. Breeding these strains together and performing genomic and metabolomic profiling revealed gene–transcript–metabolite and gene–metabolite–transcript networks. Specifically, glutamate/glutamine (Glx) was significantly correlated to argininosuccinate synthetase 1 (Ass1), arginase 1 (Arg1), phosphoenolpyruvate carboxykinase 1 (Pck1), isovaleryl coenzyme A dehydrogenase (Ivd), and alanine:glyoxylate aminotransferase (Agxt) mRNAs.[25] This is consistent with network models showing that quantitative trait loci (QTL) regulate Glx, which then regulates gene expression, or, conversely, QTL regulate mRNA abundance of the four transcripts, which then regulate Glx. These studies have the potential to uncover metabolic networks involved in the pathogenesis of diabetes.[25]

Using omics to profile mechanisms for cardiovascular disease

Sixty percent of diabetics die from cardiovascular disease (CVD), and there is a four-fold increase in morbidity and mortality from atherosclerosis in the T2DM versus non DM population.[26] Both Newgard and Domenico Accili (Columbia University) provided insight into the link between CVD and diabetes.

Why do some people with coronary artery disease experience cardiovascular events while others do not? To begin to address this question, Newgard turned to targeted metabolomic profiling. Elevated levels of short-chain dicarboxylacylcarnitines, ketone-related metabolites, and short-chain acylcarnitines were predictive of a composite endpoint of myocardial infarction (MI), repeat revascularization, or death at any point after coronary artery bypass grafting (CABG).[27] A previous study had also shown that short-chain dicarboxyla-cylcarnitines were independently associated with a greater risk of death and incidence of MI for those undergoing cardiac catheterization.[28]

Dicarboxyl acyl carnitines seem to be predictive of subsequent CV events, and therefore Newgard and colleagues are now undertaking a two-pronged approach to further characterize the association of these metabolites with cardiovascular events. Using a human genetic approach, Newgard is

performing metabolomic and genomic profiling of patients in Duke University's CATHGEN biorepository, which contains DNA and serum samples from people undergoing cardiac catheterization. To date, Newgard has profiled approximately 3500 individuals, 70% of whom have coronary artery disease and 30% of whom have diabetes. The genomic and metabolomic profiles of these patients have implicated genes involved in endoplasmic reticulum stress and in the unfolded protein response pathway. These genes are believed to play a role in modulating the concentrations of the small chain acyl carnitines.

Coordinate regulation of cholesterol, bile acid, and lipid homeostasis via FoxO–FXR interactions

The crucible of interaction between diabetes and lipoprotein metabolism may be the liver, and the Accili laboratory has dissected biochemical pathways in liver that are regulated by nutrient and insulin signaling, dependent on the action of FoxO transcription factors. FoxO was previously thought of as a modulator of hepatic glucose production solely, and Accili presented new evidence linking dysregulation of FoxO action, which can stem from hepatic insulin resistance, to dysregulation of bile acid synthesis, leading to dysregulation of cholesterol synthesis and absorption and triglyceride synthesis, all of which can affect lipoproteins associated with an increased risk for cardiovascular disease. Bile acids are synthesized from cholesterol and have feedback effects that increase lipid and cholesterol absorption, lower plasma glucose, and decrease TG synthesis and levels. Bile acids act through their own subclass of orphan nuclear receptors FXRs, and the G protein–coupled bile acid receptor, TGR5. Bile acids are synthesized from cholesterol through both classical and alternative pathways. In the alternative pathway, the side chain oxidation of cholesterol precedes the steroid ring modifications, first yielding 24-, 25-, and 27-hydroxycholesterol metabolites, opposite to the process in the classical pathway. The alternative and classical pathway bile acids share the primary bile acid chenodeoxycholic acid, with 12α hydroxylation of chenodeoxycholic acid via CYP8B1 to cholic acid. Modifications of bile acids can affect their properties and their ability to activate these receptors. In mice lacking liver FoxO1 (L-FoxO1), 12-hydroxylated bile acids are reduced, while non-12-hydroxylated bile acids are increased. This is due to a sharp reduction in the expression of *Cyp8b1*, the gene encoding the 12-hydroxylase, in L-FoxO1 mice. The increase in hydrophilic (non-12-hydroxylated) over hydrophobic (12-hydroxylated) bile acids was shown to downregulate FXR, resulting in an increase in TG synthesis. The increase in hydrophilicity of the bile acid pool also contributed to an increase in cholesterol synthesis in L-FoxO1 mice, presumably a response to its contribution to low cholesterol absorption. This metabolomics analysis led to pursuing a testable hypothesis, would administration of a FXR agonist reverse the hypertriglyceridemia of L-FoxO1 mice?

L-FoxO1 and double mutant L-FoxO1:LDLR$^{-/-}$ mice were used to test the role of FXR. Both mouse models were shown to have increased liver weight and TG content, and hypertriglyceridemia. When FXR ligand (GW4064 or cholic acid) is given to L-FoxO1 and to L-FoxO1:LDLR$^{-/-}$ mice on a cholesterol-rich western diet for 8 to 10 weeks, FXR activation with cholic acid or GW4064 prevents liver TG accumulation in both mice strains, providing mechanistic evidence of the involvement of FXR.[29]

In summary, insulin, via FoxO, regulates the balance between 12-hydroxylated and non-12-hydroxylated bile acids. When non-12-hydroxylated bile acids are predominant in mice there is decreased FXR activation increased cholesterol, TG, and FFA synthesis and increased SREBP2 activation (Fig. 4). Based on these findings, Accili speculated that dyslipidemia in diabetes could be treated by targeting components of the bile acid synthetic pathway or by providing missing bile acids.

Using omics to unravel the link between diabetes and the central nervous system

Both Accili and Gabriele Ronnett (Johns Hopkins University School of Medicine) discussed how metabolism in the brain affects food intake, energy utilization, and insulin sensitivity. The hormones insulin and leptin activate signaling pathways in the brain that decrease food intake and increase energy expenditure. However, under conditions of insulin resistance, these pathways are dysregulated. Studying how the brain regulates appetite and energy utilization identified candidate drug-susceptible targets that may be able to modify these processes during the course of diabetes.

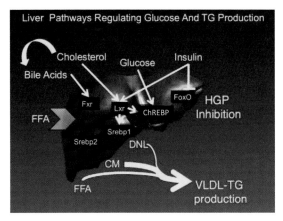

Figure 4. Outline of a working hypothesis for liver signaling pathways that affect diabetic dyslipidemia and hyperglycemia, linking bile acid, cholesterol, glucose, and insulin signaling to glucose, cholesterol, fatty acid, and triglyceride synthesis. FFA, free fatty acids; Srebp, Sterol regulatory element-binding protein; ChREBP, Carbohydrate-responsive element-binding protein; Lxr, Liver X receptor; Fxr, farnesoid X receptor; FoxO, Forkhead box O transcription factor; HGP, hepatic glucose production; DNL, *de novo* lipogenesis; TG, triglyceride; VLDL, very low density lipoprotein; CM, chylomicron. Image courtesy of Domenico Accili.

The role of FoxO1 in regulating appetite control in the brain

In the brain there are two competing populations of neurons, those that make proopiomelanocortin (POMC) and those that make neuropeptide Y/Agouti-related peptide (NPY/AgRP), that compete for regulation of energy expenditure, food intake, and satiety. By activation of the catabolic POMC neurons, insulin and leptin decrease food intake and increase energy expenditure and physical activity. Such catabolic POMC neuron activity occurs concurrently with inhibition of the anabolic AgRP neurons, which, when activated, function to increase food intake and decrease energy expenditure and physical activity. Attempts to identify drugs that can modulate POMC and AgRP neurons have been fraught with difficulty. Genetic knockouts examining the role of these two subpopulations have been generally uninformative; for example, knockout of the insulin or leptin receptor in POMC or AgRP neurons has no apparent effect on food intake. FoxO1 is a shared mediator of both pathways and its inhibition is required to induce satiety. Fasting promotes FoxO1 nuclear localization in AgRP neurons, and whereas FoxO1 is excluded from the nucleus in the fed state. Accordingly, FoxO1 ablation in AgRP neurons of mice results in reduced food intake, leanness, improved glucose homeostasis, and increased sensitivity to insulin and leptin (Fig. 5). Peripherally, there is browning of white adipocytes in AgRP FoxO1 KO mice and evidence of increased mitochondrial size/mass (EM), along with increased expression of the mitochondrial uncoupling factor UCP1 in AgRP FoxO1 KO adipocytes. Importantly, knocking out FoxO1 from AgRP neurons increases the rate of glucose disposal and decreases HGP.

Integrated omic studies have been performed to identify the FoxO1 target in AgRP neurons. Signaling activation studies show pSTAT3 (a surrogate marker of leptin activation) is decreased and pAkt is increased in AgRP FoxO1 KO neurons. Immunohistochemistry showed pS6 signaling is increased in AgRP FoxO1 KO neurons projecting from the arcuate nucleus, signaling a state of abundant nutrients.[30]

Transcriptomic and electrical excitability studies indicate that AgRP FoxO1 KO neurons are less excited/more inhibited. Patch clamping shows FoxO1 KO AgRP neurons are constitutively inhibited. Expression profiling of flow-sorted FoxO1-deficient AgRP neurons identified an increase in GABA receptor (inhibitory) expression and a decrease in glutamate receptor (excitatory) expression, as well as the G protein–coupled receptor Gpr17 as a FoxO1 target whose expression is regulated by nutritional status (Fig. 5). Intracerebroventricular injection of Gpr17 agonists induces food intake, whereas the Gpr17 antagonist cangrelor curtails it. These effects are absent in AgRP-FoxO1 knockouts, suggesting that pharmacological modulation of this pathway, perhaps with brain-permeable Gpr17 agents, has therapeutic potential to treat obesity.[30]

The role of fatty acid metabolism in the regulation of energy balance

The brain is a highly metabolic organ capable of fatty acid oxidation and storage, and the focus of Ronnett's group is the investigation of the hypothesis that pharmacological alteration of fatty acid flux can alter food intake. Ronnett's lab has focused on three metabolic enzyme candidates, key for the accumulation of long chain fatty acids, as targets for obesity intervention (Fig. 6). Fatty acid synthase (FAS) is a lipogenic enzyme that generates saturated long-chain fatty acids such as palmitate, which

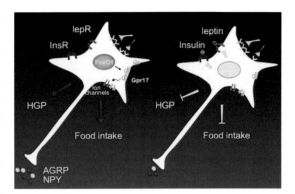

Figure 5. Model for regulation of food intake and hepatic glucose activity by FoxO1 and Gpr17. The G protein–coupled receptor Gpr17 is a FoxO1 target whose expression is regulated by nutritional status, and may play a role in mediating food intake. InsR, insulin receptor; lepR, leptin receptor; HGP, hepatic glucose production; NPY/AGRP, neuropeptide Y/Agouti-related peptide. Image courtesy of Domenico Accili.

has 16 carbons. Carnitine palmitoyl-transferase-1 (CPT-1 isoforms) is requisite for the entry of long-chain fatty acids into mitochondria for oxidation. Glycerol-3-phosphate acyltransferases (GPATs) catalyze the first and rate-limiting step for fatty acids to phospholipid and triglyceride syntheses. In general, increased fatty acid oxidation is characteristic of the fasted state, and Ronnett hypothesized that either FAS or GPAT inhibition, or CPT-1 stimulation in the central nervous system, would decrease food intake and body weight. Ronnett tested this hypothesis using three small molecules: C75, an inhibitor of FAS and activator of CPT-1; FSG67, an inhibitor of GPAT; and C89b, an activator of CPT-1; and confirmed that these molecules reduce food intake, increase energy expenditure, and enhance fatty acid oxidation to decrease adiposity and body weight.[31–34]

Central administration of these compounds alters neuronal activity in select hypothalamic nuclei that control food intake and energy expenditure. In these hypothalamic nuclei, compound treatment leads to altered gene expression and production of neuropeptides germane to energy balance, consistent with homeostatic responses to CNS perception of physiologically positive energy balance.

Recent work has shown that obeseogenic diets high in saturated fatty acids, known to cause peripheral inflammation and exacerbate diabesity, also induce CNS inflammation, ER stress, and oxidative

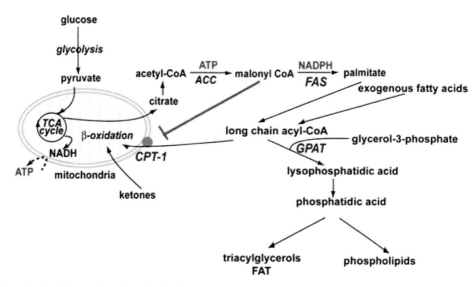

Figure 6. The role of fatty acid synthase (FAS), glycerol-3-phosphate acyltransferase (GPAT), and carnitine palmitoyl transferase 1 (CPT-1) in brain fatty acid metabolism. C75, an inhibitor of FAS and activator of CPT-1, FSG67, an inhibitor of GPAT, and C89b, an activator of CPT-1, all reduce food intake, increase energy expenditure, and enhance fatty acid oxidation to decrease adiposity and body weight. Image courtesy of Gabriele Ronnett.

stress, and that this may contribute to the development of metabolic syndrome.

Both C75 and FSG67 induced weight loss in obese mice. Examining the genetic effects of these compounds revealed that the synthesis of enzymes involved in fatty acid storage was downregulated, whereas the synthesis of enzymes involved in fat disposition was upregulated. FSG67 is currently in preclinical safety tests. C89b, the CPT-1 stimulator, also decreased food intake and induced weight loss, consistent with the role of CPT-1 in promoting lipid oxidation. The effects seen with C89b in mice, however, were more dramatic and longer-lasting than those seen with C75 or with FSG67.

To investigate the mechanisms of these compounds' effects, researchers are conducting metabolomic studies in neurons *in vitro*. So far, they have seen that C75 and FSG67 increase reactive oxygen species while reducing the secretion of inflammatory cytokines. Based on these data, Ronnett speculated that C75 and FSG67 are not just altering fatty acid metabolism in the neurons but may also have long-term effects on inflammation in the brain. To understand what other pathways are affected by alteration of fatty acid flux and to elucidate what metabolic changes are affecting the observed changes in inflammatory signals, Ronnett is undertaking a full metabolomic profile in primary hy-

pothalamic and cortical neurons treated with palmitate, C75, and FSG67. Initial results indicate that primary hypothalamic neurons show different responses to these agents. Under normal nutrient conditions, hypothalamic neurons did not have a significant fatty acid profile response to C75; however, in the setting of nutrient (palmitate) excess, C75 did have a significant effect and caused a decrease di- and triglycerides in primary hypothalamic neurons. FAS inhibition did increase TCA metabolite levels, which suggests a pathway for modification of ATP levels other than manipulation of fatty acid oxidation, which may be part of an AMPK related mechanism of action for these agents.[35]

Targeted versus global untargeted metabolomics profiling as a tool for metabolic phenotyping

An industry panel—Steven Fischer (Agilent Technologies), Suma Ramagiri (AB SCIEX), John Ryals (Metabolon), Mark Sanders (Thermo Fisher Scientific), John Shockcor (Waters Corporation), and Joe Shambaugh (Genedata)—was asked to discuss the benefits, drawbacks, and areas of future development for targeted versus global untargeted profiling as tools for metabolic phenotyping.

In general, mastering the tools of chromatographic separation methods takes precedence over metabolite identification. Normal phase and reversed phase chromatography have synergism for global small molecule separation and identification.[36] Supercriticial fluid chromatography, for example, may provide a new modality for future global lipid analysis.[37] Derivatization can aid targeted LC/MS/MS analysis, as used in amino acid[38,39] and acyl carnitine analysis.[40,41]

A targeted quantitative approach, using GC/MS and LC/MS/MS, is the best first approach for any metabolomics/lipidomics problem. This should be followed with a global profiling paradigm, first aimed at getting the best possible exact MS data, in particular with retention time locked databases, subsequently re-run to obtain MS/MS to aid database searching. The principal challenges in global profiling are the creative use of algorithms for the separation of peaks from noise, optimal data mining paradigms and databases, and for biofluids determining the source for the metabolites identified[42–47] (Fig. 7).

Metabolite biomarkers include those synthesized *in vivo* and those derived from exogenous sources, including the microbiome. The consensus was that humans are capable of synthesizing roughly 2500 compounds. As reviewed in Dunn *et al.*,[48] 2,000–7,000 metabolic features can be detected in a serum or plasma sample. A single metabolite can be detected as different ion types: for example, as protonated and deprotonated ions, adduct ions, isotopomers, fragment ions, dimers, and trimers. Therefore, a large number of metabolic features identified correspond to a smaller number of actual metabolites.[49] Humans may contain more molecules than they are able to directly synthesize, due to microbiome metabolism, drugs, or dietary supplements. Differences in the amount of compounds in human plasma found at different facilities stem, in part, from whether pooled human samples were used versus individual test subjects, as well as some differences due to the particular MS platform used. Pooled plasma samples have as many as 2000 compounds (Fischer, private communication), while individual subjects have at least 500–600 compounds.[50]

The dataset derived from untargeted mass spectrum analysis may be very noisy, with noise in unit mass and/or accurate mass instruments being ~80% of the total data collected.[51] Optimal peak identification/separation of sample peaks

from chemical noise, and clustering of their GC/MS and LC/MS data before library search for metabolite identification, is facilitated by software packages such as Mass Profiler Professional, Thermo Scientific Sieve,[52] Genedata Expressionist for Mass Spec,[53] Transomics, and XC/MS[45] (see Fig. 7).

The current data mining paradigm involves extracting data using a naive feature extractor and performing compound identification on the reconstructed spectra. Untargeted mass spectrum analysis is facilitated by assembly of a database composed of a large number of library standards. Each standard entry can have a number of features, such as a retention time index, MS spectra, and MS/MS fragmentation spectra, obtained at different collision energies. Retention time libraries can be machine- and column-specific, as different machines have different sensitivities, and some problems requiring nano-UPLC will necessarily have a different retention time library than standard UPLC. As mentioned, due to the redundancy of the ion spectra, each library entry may have ~ 10 or more features, as each molecular standard can be associated with 5–10 ion features.[48,51,54] The current Agilent-METLIN database and MS/MS library contains ~ 45,000 compounds, with ~ 9000 compounds having MS/MS spectra.[44] METLIN data has been acquired using a collision cell shared by triple quadrupole and qTOF machines. MS/MS spectra are collected in both positive- and negative-ion mode and at 10, 20, and 40 eV collision energies. Those spectra that have at least one ion with ~ 1000 counts of signal are retained for entry into the MS/MS library. The spectra are edited to only include ion signals coming from the standard, and the reported mass is corrected to its theoretical mass.[44] GC/MS metabolite identifications are facilitated by well-defined MS conditions and libraries, as reviewed Kind and Fiehn,[55] and METLIN, Mass Frontier, and m/z Cloud[a] are establishing databases that together cover a wide variety of MS platforms.

The loose fit of MS[n] spectra with the METLIN database suggests that MS and MS[n] spectra generated on LTQ-Orbitrap machines are best identified by Mass Frontier.[55] The larger the database, the better it works, and the m/z Cloud community-based effort aims to establish a comprehensive library of

---

[a]http://www.mzcloud.org/

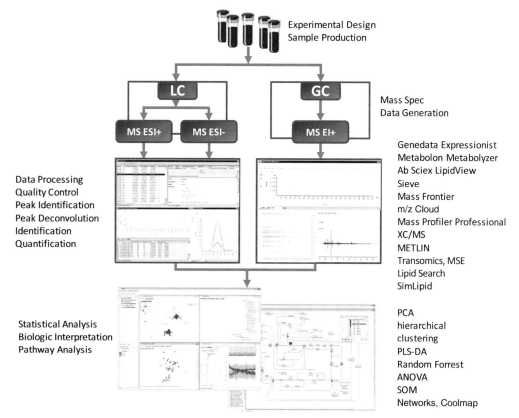

Figure 7. Overview of metabolomic data generation and data analysis. The flowchart used for metabolite extraction, data mining, and metabolite identification is detailed. This illustrates sample preparation, mass spectrometric analysis, peak extraction/identification and compound quantification, and statistical data analysis for biomarker identification and mapping of biomarkers to metabolic pathways. Samples, in general, can be divided into separate groups for gas chromatography/ mass spectrometry (GC/MS) and liquid chromatography/MS (LC/MS). LC/MS is further divided to examine both positively and negatively charged ions, first done with full scan for exact mass, and after with LC/MS/MS for identification with databases such as METLIN, SimLipid, LipidView, Lipid Search or Mass Frontier, or in the case of GC/MS, Fiehn and NIST libraries. *Data preprocessing* covers the software programs that process complex raw data to clean data. Data preprocessing programs (for example, Genedata Expressionist for Mass Spec, Transomics, XC/MS, Sieve, Metabolon Metabolyzer) are used to separate peaks from noise, and then database searching can be accomplished. Genedata Expressionist *for Mass Spec*, Transomics, and XC/MS are mass spectrometer platform–independent. A variety of techniques can be used for statistical analysis, including principal component analysis (PCA), partial least squares discriminant analysis (PLS-DA), analysis of variance (ANOVA), Random Forrest, self-organizing maps (SOM), and platform-independent software such as SIMCA-P, Transomics and Genedata and Expressionist *for Mass Spec* can be used for such analyses. An overview of such statistical methods can be found in Madsen *et al.*[57] Image courtesy of Joe Shambaugh.

high quality spectral trees to improve the structural elucidation of unknowns by identifying compounds even when they are not present in the library, using spectral tree searches. For example, individual MS[n] spectra can be searched against the m/z Cloud library to retrieve structural or substructural hits. The challenge is reassembly, which can be expert-motivated and have input from correlations with other metabolites to assemble the puzzle.[55]

Lipidomic database searches benefit from the LipidMaps initiative,[56] which has resulted in dedi-

cated commercially available *in silico* lipid databases such as LipidView (Ab Sciex), SimLipid (Biosoft), or Lipid Search (MKI), enabling one to uniquely identify over 20,000 lipid species using characteristic lipid fragments.[50]

The use of pathways as a means to interpret metabolomics data acquired using non-targeted data acquisition strategies opens up a different approach to data mining. By using pathways for biological interpretation, the researcher has defined the metabolites in the pathway(s) as a target compound list. The identified target list then can be used for statistical analysis[57] (Fig. 7) rather than just analyzing features. This compound list can be used as the template for further mining the pathway(s) using targeted identification and data extraction.

Future developmental work could center on matching possible metabolites at successive nodes, integrating searching with pathway databases for both GC and LC. For GC, this would involve theoretical calculations of derivatization effects.[58] As compounds are actually identified, a database can be created that records this information for future use in compound identification. Another possibility is to use genome-wide association studies (GWAS)[59] data to see if there is an association to the molecule of interest. At times, associating a particular allele to specific metabolite biomarkers may suggest a known gene or a gene of a known class.

An unknown compound can be identified, tracked, and quantitated with relative or semi-quantification even though its true identity is not known. If such a molecule becomes an important biomarker, there are several approaches that can be used to either suggest an identity or get clues as to the identity. Biochemicals are typically not independent variables; they change in groups that are related biosynthetically or functionally, and statistical correlative methods can be of use to postulate relationships. Important biomarkers identified in this manner can have their mass accurately determined, atomic composition calculated, and identification made more complete by using MS^n analysis. Such approaches can give scientists better ideas about the identity of the metabolite, its molecular composition, and the pathways involved in its metabolism.

Conclusion

Metabolomics, lipidomics and fluxomics technologies are still in their relative infancy for general biomarker discovery, and can be integrated with other omics (proteomic, transcriptomic and genomic) to reveal a more complete picture of diabesity disease mechanisms. Complementary approaches to multi-omic metabolic pathway analysis may involve a tiered hypothesis-driven framework, to determine whether another omic may be indicated. Additionally, emerging visualization tools for shotgun omic data evaluations allow the generation of hypotheses about the metabolic networks that respond to particular types of intervention.

Fluxomics and both targeted and global untargeted metabolomics profiling can be used, in conjunction with mouse dietary and genetic models, and human clinical studies, to unravel the link(s) between diabetes and obesity, and to profile metabolic mechanisms in and between the CNS and periphery (liver, fat, muscle) that may affect plasma metabolic biomarkers. In general, metabolomics, lipidomics and fluxomics hold the promise not only for diagnostic evaluation in routine clinical use to predict disease progression and outcomes, but also, by nature of their pathophysiological relevance, for identification of target pathways that may relate to molecular mechanisms. Identified pathways then can be the focus of drug development for future use in therapeutics for personalized medicine.

Acknowledgements

The conference "Application of Combined 'omics Platforms to Accelerate Biomedical Discovery in Diabesity" was presented by Hot Topics in Life Sciences and the Sackler Institute for Nutrition Science at the New York Academy of Sciences.

Cross-omic evaluation methodology development was supported by 2R01DK058132 (IJK), 3R37DK058282 (IJK and DA), P60DK020541 (Einstein DRTC), and 1U19AI091175 (Einstein CMCR).

Conflicts of interest

The authors declare no conflicts of interest.

References

1. Centers for Disease Control. 2011. Number of Americans with diabetes rises to nearly 26 million. More than a third of adults estimated to have prediabetes. January 26, 2011. Cited March 12. 2013. http://www.cdc.gov/media/releases/2011/p0126_diabetes.html.
2. Kelley, D.E. 2005. Skeletal muscle fat oxidation: timing and flexibility are everything. *J. Clin. Invest.* **115:** 1699–1702.
3. Galgani, J.E., C. Moro & E. Ravussin. 2008. Metabolic flexibility and insulin resistance. *Am. J. Physiol. Endocrinol. Metab.* **295:** E1009–1017.
4. Farag, Y.M. & M.R. Gaballa. 2011. Diabesity: an overview of a rising epidemic. *Nephrol. Dial. Transplant.* **26:** 28–35.
5. Kaufman, F.R. 2005. The Obesity-Diabetes Epidemic That Threatens America—And What We Must Do to Stop It. Random House Digital.

[This article was corrected on 23 May 2013 after original online publication. Additional Acknowledgements were added.]

6. Xu J., L. Gowen, C. Raphalides, K.K. Hoyer, *et al.* 1006. Decreased hepatic futile cycling compensates for increased glucose disposal in the Pten heterodeficient mouse. *Diabetes.* **55:** 3372–3380.

7. Vaitheesvaran B., F.Y. Chueh, J. Xu, *et al.* 2010. Advantages of dynamic "closed loop" stable isotope flux phenotyping over static "open loop" clamps in detecting silent genetic and dietary phenotypes. *Metabolomics.* **6:** 180–190.

8. Vaitheesvaran B., L. Yang, K. Hartil,*et al.* 2012. Peripheral effects of FAAH deficiency on fuel and energy homeostasis: role of dysregulated lysine acetylation. *PLoS One.* **7:** e33717.

9. Xu J., G. Xiao, C. Trujillo, *et al.* 2002. Peroxisome proliferator-activated receptor alpha (PPARalpha) influences substrate utilization for hepatic glucose production. *J. Biol. Chem.* **277:** 50237–50244.

10. Vaitheesvaran B., D. LeRoith, I.J. Kurland. 2010. MKR mice have increased dynamic glucose disposal despite metabolic inflexibility, and hepatic and peripheral insulin insensitivity. *Diabetologia.* **53:** 2224–2232.

11. Xu J., V. Chang, S.B. Joseph, *et al.* 2004. Peroxisomal proliferator-activated receptor alpha deficiency diminishes insulin-responsiveness of gluconeogenic/glycolytic/pentose gene expression and substrate cycle flux. *Endocrinology.* **145:** 1087–1095.

12. Pei L., H. Waki, B. Vaitheesvaran, D.C. Wilpitz, *et al.* 2006. NR4A orphan nuclear receptors are transcriptional regulators of hepatic glucose metabolism. *Nat. Med.* **12:** 1048–1055.

13. Xu J., W.N. Lee, J. Phan, M.F. Saad, *et al.* 2006. Lipin deficiency impairs diurnal metabolic fuel switching. *Diabetes.* **55:** 3429–3438.

14. Xu J., W.N. Lee, G. Xiao, *et al.* 2003. Determination of a glucose-dependent futile recycling rate constant from an intraperitoneal glucose tolerance test. *Anal. Biochem.* **315:** 238–246.

15. Yang L., B. Vaitheesvaran, K. Hartil, A.J. Robinson, *et al.* 2011. The fasted/fed mouse metabolic acetylome: N6-acetylation differences suggest acetylation coordinates organ-specific fuel switching. *J. Proteome Res.* **10:** 4134–4149.

16. Zhong L., A. D'Urso, D. Toiber, *et al.* 2010. The histone deacetylase Sirt6 regulates glucose homeostasis via Hif1alpha. *Cell.* **140:** 280–293.

17. Zong H., C.C. Wang, B. Vaitheesvaran, *et al.* 2011. Enhanced Energy Expenditure, Glucose Utilization And Insulin Sensitivity In VAMP8 Null Mice. *Diabetes.* **60:** 30–38.

18. Karnovsky A., T. Weymouth, T. Hull, *et al.* 2012. Metscape 2 bioinformatics tool for the analysis and visualization of metabolomics and gene expression data. *Bioinformatics.* **28:** 373–380.

19. Newgard C.B., J. An, J.R. Bain, *et al.* 2009. A branched-chain amino acid-related metabolic signature that differentiates obese and lean humans and contributes to insulin resistance. *Cell Metab.* **9:** 311–326.

20. Huffman K.M., S.H. Shah, R.D. Stevens, *et al.* 2009. Relationships between circulating metabolic intermediates and insulin action in overweight to obese, inactive men and women. *Diabetes Care.* **32:** 1678–1683.

21. Tai E.S., M.L. Tan, R.D. Stevens, *et al.* 2010. Insulin resistance is associated with a metabolic profile of altered protein metabolism in Chinese and Asian-Indian men. *Diabetologia.* **53:** 757–767.

22. Shah S.H., D.R. Crosslin, C.S. Haynes, *et al.* 2012. Branched-chain amino acid levels are associated with improvement in insulin resistance with weight loss. *Diabetologia.* **55:** 321–330.

23. Laferrere B., D. Reilly, S. Arias, *et al.* 2011. Differential metabolic impact of gastric bypass surgery versus dietary intervention in obese diabetic subjects despite identical weight loss. *Sci. Transl. Med.* **3:** 80re2.

24. Herman M.A., P. She, O.D. Peroni, *et al.* 2010. Adipose tissue branched chain amino acid (BCAA) metabolism modulates circulating BCAA levels. *J. Biol. Chem.* **285:** 11348–11356.

25. Ferrara C.T., P. Wang, E.C. Neto, *et al.* 2008. Genetic networks of liver metabolism revealed by integration of metabolic and transcriptional profiling. *PLoS Genet.* **4:** e1000034.

26. Tan M.H. 1999. Diabetes and coronary artery disease. *Diabetes Spectr.* **12:** 80–83.

27. Shah A.A., D.M. Craig, J.K. Sebek, *et al.* 2012. Metabolic profiles predict adverse events after coronary artery bypass grafting. *J. Thorac. Cardiovasc. Surg.* **143:** 873–878.

28. Shah S.H., J.R. Bain, M.J. Muehlbauer, *et al.* 2010. Association of a peripheral blood metabolic profile with coronary artery disease and risk of subsequent cardiovascular events. *Circ. Cardiovasc. Genet.* **3:** 207–214.

29. Haeusler R.A., M. Pratt-Hyatt, C.L. Welch, *et al.* 2012. Impaired generation of 12-hydroxylated bile acids links hepatic insulin signaling with dyslipidemia. *Cell Metab.* **15:** 65–74.

30. Ren H., I.J. Orozco, Y. Su, *et al.* 2012. FoxO1 target Gpr17 activates AgRP neurons to regulate food intake. *Cell.* **149:** 1314–1326.

31. Kim E.K., I. Miller, S. Aja, *et al.* 2004. C75, a fatty acid synthase inhibitor, reduces food intake via hypothalamic AMP-activated protein kinase. *J.Biol. Chem.* **279:** 19970–19976.

32. Tu Y., J.N. Thupari, E.K. Kim, *et al.* 2005. C75 alters central and peripheral gene expression to reduce food intake and increase energy expenditure. *Endocrinology.* **146:** 486–493.

33. Aja S., L.E. Landree, A.M. Kleman, *et al.* 2008. Pharmacological stimulation of brain carnitine palmitoyl-transferase-1 decreases food intake and body weight. *Am. J. Physiol.* **294:** R352–361.

34. Kuhajda F.P., S. Aja, Y. Tu, *et al.* 2011. Pharmacological glycerol-3-phosphate acyltransferase inhibition decreases food intake and adiposity and increases insulin sensitivity in diet-induced obesity. *Am. J. Physiol.* **301:** R116–130.

35. Kleman A.M., J.Y. Yuan, S. Aja, *et al.* 2008. Physiological glucose is critical for optimized neuronal viability and AMPK responsiveness in vitro. *J. Neurosci. Methods.* **167:** 292–301.

36. Patti G.J. 2011. Separation strategies for untargeted metabolomics. *J. Sep. Sci.* **34:** 3460–3569.

37. Bamba T., J.W. Lee, A. Matsubara & E. Fukusaki. 2012. Metabolic profiling of lipids by supercritical fluid chromatography/mass spectrometry. *J. Chromatogr.* **1250:** 212–219.

38. White J.A., R.J. Hart & J.C. Fry. 1986. An evaluation of the Waters Pico-Tag system for the amino-acid analysis of food materials. *J. Automat. Chem.* **8:** 170–177.

39. Salazar C., J.M. Armenta, D.F. Cortes & V. Shulaev. 2012. Combination of an AccQ.Tag-ultra performance liquid chromatographic method with tandem mass spectrometry for the analysis of amino acids. *Methods Mol. Biol.* **828:** 13–28.

40. van Vlies N., L. Tian, H. Overmars, *et al.* 2005. Characterization of carnitine and fatty acid metabolism in the long-chain acyl-CoA dehydrogenase-deficient mouse. *Biochem. J.* **387:** 185–193.

41. Vreken P., A.E.M Van Lint, A.H. Bootsma, *et al.* 1999. Quantitative plasma acylcarnitine analysis using electrospray tandem mass spectrometry for the diagnosis of organic acidaemias and fatty acid oxidation defects. *J. Inher. Metab. Dis.* **22:** 302–306.

42. Kind T., G. Wohlgemuth, Y. Lee do, *et al.* 2009. FiehnLib: mass spectral and retention index libraries for metabolomics based on quadrupole and time-of-flight gas chromatography/mass spectrometry. *Anal. Chem.* **81:** 10038–10048.

43. Lu H., W.B. Dunn, H. Shen, *et al.* 2008. Comparative evaluation of software for deconvolution of metabolomics data based on GC-TOF-MS. *Trends Anal. Chem.* **27:** 215–227

44. Tautenhahn R., K. Cho, W. Uritboonthai, *et al.* 2012. An accelerated workflow for untargeted metabolomics using the METLIN database. *Nat. Biotech.* **30:** 826–828.

45. Tautenhahn R., G.J. Patti, D. Rinehart & G. Siuzdak. 2012. XCMS Online: a web-based platform to process untargeted metabolomic data. *Anal. Chem.* **84:** 5035–5039.

46. Dehaven C.D., A.M. Evans, H. Dai & K.A. Lawton. Organization of GC/MS and LC/MS metabolomics data into chemical libraries. *J. Cheminform.* **2:** 9.

47. Gürdeniz G., M. Kristensen, T. Skov & L.O. Dragsted. 2012. The Effect of LC-MS Data Preprocessing Methods on the Selection of Plasma Biomarkers in Fed vs. Fasted Rats. *Metabolites.* **2:** 77–99.

48. Dunn W.B., D. Broadhurst, P. Begley, *et al.* 2011. Procedures for large-scale metabolic profiling of serum and plasma using gas chromatography and liquid chromatography coupled to mass spectrometry. *Nat Protoc.* **6:** 1060–1083.

49. Dehaven C.D., A.M. Evans, H. Dai & K.A. Lawton. 2010. Organization of GC/MS and LC/MS metabolomics data into chemical libraries. *J. Cheminform.* **2:** 9.

50. Simons B., D. Kauhanen, T. Sylvänne, *et al.* 2012. Shotgun Lipidomics by Sequential Precursor Ion Fragmentation on a Hybrid Quadrupole Time-of-Flight Mass Spectrometer. *Metabolites.* **2:** 195–213.

51. Evans AM, H. Dai & C.D. DeHaven. 2012. Categorizing Ion -Features in Liquid Chromatography/Mass Spectrometry Metobolomics Data. *Metabolomics.* **2:** 110.

52. Athanas M., D.A. Peake, M. Dreyer, *et al.* 2012. Applying Q Exactive Benchtop Orbitrap LC-MS/MS and SIEVE 2.0 Software for Cutting-Edge Metabolomics and Lipidomics Research. 2012. Cited March 12, 2013. https://static.thermoscientific.com/images/D20896~.pdf

53. Hoefkens J., T. Kind, K. Pinkerton & O. Fiehn. 2008. An automated workflow for rapid alignment and identification of lipid biomarkers obtained from chip-based direct infusion nanoelectrospray tandem mass spectrometry. 2008. Cited March 12, 2013. http://www.genedata.com/fileadmin/docu ments/Landing/Pages/Mass/Spec/Center/2008/Poster/ASMS /UC/Davis.pdf

54. Evans A.M., C.D. DeHaven, T. Barrett, *et al.* 2009. Integrated, nontargeted ultrahigh performance liquid chromatography/electrospray ionization tandem mass spectrometry platform for the identification and relative quantification of the small-molecule complement of biological systems. *Anal. Chem.* **81:** 6656–6667.

55. Kind T.F. & O. Fiehn. 2010. Advances in structure elucidation of small molecules using mass spectrometry. *Bioanal. Rev.* **2:** 23–60.

56. Fahy E., D. Cotter, M. Sud & S. Subramaniam. 2011. Lipid classification, structures and tools. *Biochim. Biophys. Acta.* **1811:** 637–647.

57. Madsen R., T. Lundstedt & J. Trygg. 2010. Chemometrics in metabolomics–a review in human disease diagnosis. *Anal Chim. Acta.* **659:** 23–33.

58. Kumari S., D. Stevens, T. Kind, *et al.* 2011. Applying in-silico retention index and mass spectra matching for identification of unknown metabolites in accurate mass GC-TOF mass spectrometry. *Anal. Chem.* **83:** 5895–5902.

59. Hindorff L.A., P. Sethupathy, H.A. Junkins, *et al.* 2009. Potential etiologic and functional implications of genome-wide association loci for human diseases and traits. *Proc. Natl. Acad. of Sci.* **106:** 9362–9367.

Appendix

Additional reading

Newgard C.B. 2012. Interplay between lipids and branched-chain amino acids in development of insulin resistance. *Cell Metab.* **15:** 606–614

Sana T.R., K. Waddell & S. M. Fischer. 2008. A sample extraction and chromatographic strategy for increasing LC/MS detection coverage of the erythrocyte metabolome. *J. Chromatogr. B.* **871:** 314–321

Sana T.R., J.C. Roark, X. Li, *et al.* 2008. Molecular Formula and METLIN Personal Metabolite Database Matching Applied to the Identification of Compounds Generated by LC/TOF-MS. *J. Biomol. Tech.* **19:** 258–266

Adamski J. & K. Suhre. 2013. Metabolomics platforms for genome wide association studies-linking the genome to the metabolome. *Curr. Opin. Biotechnol.* **24:** 39–47

Ann. N.Y. Acad. Sci. ISSN 0077-8923

Prioritizing health disparities in medical education to improve care

Temitope Awosogba,[1] Joseph R. Betancourt,[2,3] F. Garrett Conyers,[3] Estela S. Estapé,[4] Fritz Francois,[5] Sabrina J. Gard,[5] Arthur Kaufman,[6] Mitchell R. Lunn,[3,7] Marc A. Nivet,[8] Joel D. Oppenheim,[5] Claire Pomeroy,[9] and Howa Yeung[5]

[1]Mount Sinai School of Medicine, New York, New York. [2]The Disparities Solutions Center, Massachusetts General Hospital, Boston, Massachusetts. [3]Harvard Medical School, Boston, Massachusetts. [4]Medical Sciences Campus, University of Puerto Rico, San Juan, Puerto Rico. [5]Office of Diversity Affairs, New York University School of Medicine, New York, New York. [6]University of New Mexico, Albuquerque, New Mexico. [7]Brigham and Women's Hospital, Boston, Massachusetts. [8]Association of American Medical Colleges, Washington, DC. [9]School of Medicine, University of California, Davis, Sacramento, California

Address for correspondence: annals@nyas.org

Despite yearly advances in life-saving and preventive medicine, as well as strategic approaches by governmental and social agencies and groups, significant disparities remain in health, health quality, and access to health care within the United States. The determinants of these disparities include baseline health status, race and ethnicity, culture, gender identity and expression, socioeconomic status, region or geography, sexual orientation, and age. In order to renew the commitment of the medical community to address health disparities, particularly at the medical school level, we must remind ourselves of the roles of doctors and medical schools as the gatekeepers and the value setters for medicine. Within those roles are responsibilities toward the social mission of working to eliminate health disparities. This effort will require partnerships with communities as well as with academic centers to actively develop and to implement diversity and inclusion strategies. Besides improving the diversity of trainees in the pipeline, access to health care can be improved, and awareness can be raised regarding population-based health inequalities.

Keywords: health disparities; social determinants; diversity; cross-cultural education

Introduction

Individual Americans—subject to a large number of social, cultural, economic, or geographical factors influencing their health—may find themselves in vastly different situations with regard to health outcomes and health care. For example, according to a 2011 U.S. Department of Health and Human Services report,[1] the percentage of Americans with two or more chronic health conditions correlates strongly with poverty level, and this correlation is growing. While income inequities are responsible for much of the health disparities in the United States, there are many other factors, of which only some can be readily explained. An African American child is more likely to develop asthma than a white child within the same income bracket. While obesity does not show much correlation with the highest education obtained by an individual, childhood obesity is strongly correlated with the highest education obtained by the head of the household. At any income level, Hispanic and Asian Americans are less likely to have health insurance than African Americans or non-Hispanic whites. The distribution of doctors across the United States is far from uniform: whereas there are 40 doctors for every 10,000 people in Massachusetts, there are 17 in Idaho.

The conference "Prioritizing Health Disparities in Medical Education to Improve Care", held on October 2, 2012, convened medical school faculty and administrators, educators, and students to discuss how to create real solutions at the level of medical schools. In his introductory remarks, Fitzhugh Mullan (the George Washington University) recounted the roots of the struggle against health disparities during the civil rights movement of the 1960s, and how academic study of health disparity finally attained legitimacy with the publication of the Institute of Medicine's 2003 report *Unequal*

doi: 10.1111/nyas.12117

Treatment: Confronting Racial and Ethnic Disparities in Health Care.[2]

Medical schools have long operated on the bases of research, education, and patient care. Each of these pillars suggests an ethical imperative for universities to actively pursue the social mission of disparities reduction. According to Mullan, this social mission would include mainstreaming diversity, assessing graduates' goals and career development (with an eye toward underserved communities), advancing access to care, and raising awareness of health disparities. Two distinct but complementary tactics were highlighted by numerous speakers: a process of increasing communication, outreach, and, ultimately, aligning the needs of the community with the resources of the university; and new initiatives not only to increase the numbers of underrepresented minorities entering medical education, but also to encourage those that have to stay within academic spheres of medicine.

Social determinants of health

Linking university health resources to social determinants in the community

Arthur Kaufman (University of New Mexico (UNM)) discussed means of linking university health resources to the social factors contributing to health and disease, using the example of active programs at the UNM. In the United States, health services consume over 90% of the nation's healthcare budget but contribute no more than 10–15% to the nation's health. In New Mexico, this disparity between health expenditure and health outcome can be illustrated through the mortality from diabetes among the different populations in the state. Native Americans receive some of the best screening and treatment for diabetes from the Indian Health Service and special programs targeting Native Americans with diabetes. Yet, Native Americans have the highest death rates from that disease. The high-quality care they receive cannot compensate for decades of chronic poverty, high unemployment, poor educational attainment, unhealthy food options, poor housing, and a fragmented social network—the social determinants of health and disease.

Universities can play a major role in addressing social determinants through education and clinical service. At the UNM, all medical students graduate with a Public Health Certificate, which awards 17 transferrable credits toward an MPH degree.[3] Kaufman emphasized that students not only learn about the importance of the social determinants, but also gain frontline experiences and skills in how to address them. And the UNM's investment in educating family medicine residents in rural and underserved communities, with its high rate of subsequent recruitment to those communities,[4] leads not only to better access to health service but also to a positive local economic impact. Each physician practicing in a rural community hires on the average 18 people directly and indirectly and generates approximately $1 million in business each year.

The UNM's Health Extension Rural Offices (HEROs) serve as a model for addressing social determinants. It is adapted from the U.S. Department of Agriculture's Cooperative Extension Service, which links resources from each state's land grant university to the needs of farmers and farm families.[5] HERO agents are recruited from and work in rural and urban underserved communities across the state. They link community health priorities with UNM resources and monitor the effectiveness of university programs in addressing community needs. Examples of their work include the recruitment of community preceptors of health science students, the writing of a grant to establish a federally qualified health center in a rural community with insufficient access for the uninsured, and the creation of a tele-pharmacy service for a frontier community that had lost its pharmacist.

The research agenda of the University of New Mexico has been influenced by the advent of HEROs and the recognition of social determinants. Where the Health Science Center's signature research programs responded to major NIH funding sources (cancer, cardiovascular and metabolic diseases, brain, and behavior), New Mexico's county health planning councils had a different set of health priorities (substance abuse, teen pregnancy, obesity, access to care, violence, and diabetes). Each HERO agent plays a role in linking community health questions to the research resources of the university.

HEROs further address social determinants by amplifying the impact of health systems in rural and underserved communities through the recruitment and training of community health workers (CHWs). CHWs are community residents who are culturally and linguistically competent and who, more than any other member of the health team, work with patients in addressing the social determinants of health. They improve patients' health literacy, help them navigate the health system, facilitate doctor visits, and assist in applying for financial assistance, obtaining food vouchers, and filling out housing applications. The impact of CHWs on the health team led to two Medicaid managed care companies in New Mexico employing them to improve access to primary care and reduce preventable emergency room visits, hospitalizations, and overuse of medications.[6]

Finally, to mobilize the support of all components of the Health Sciences Center toward the common purpose of institutional responsibility for community health, the institution leaders adopted a new vision, which states, "The University of New Mexico Health Sciences Center will work with Community Partners to help New Mexico make more progress in health and health equity than any other state by 2020." This vision cannot be achieved without institutional focus in all its mission areas on the underlying social determinants of health.

Health disparities and social justice: a call to action for academic health centers

A society's greatness is defined by how it treats its most vulnerable members. By this metric, Claire Pomeroy (University of California, Davis) argued, the United States has failed to achieve greatness. Indeed, Pomeroy stated, our country lacks one of the most basic components of justice that everyone deserves—health care and other services that enable each of us to live the healthiest life possible. She emphasized that now is the time for academic health centers to fully leverage their unique position at the intersection of education, research, and clinical care to lead action that advances the nation's health and well-being.

Despite the fact that the United States spends about twice as much per capita as other developed nations and nearly 18 percent of its gross domestic product (GDP) on health care, U.S. health outcomes are poor. Compared to 34 Organisation for Economic Co-operation and Development (OECD) countries, the United States ranks only 29th for men and 28th for women in life expectancy, and 31st in infant mortality.[7] Moreover, health outcomes in the United States are disproportionately distributed, with significant health disparities relative to race, ethnicity, socioeconomic status, edu-

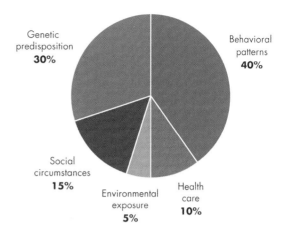

Figure 1. Proportional contributions of determinants of health to premature death.

cation, immigration status, sexual orientation, and geography.

For example, Hispanics are twice as likely as non-Hispanic whites to report poor or fair health status.[8] People with less than a high school education are more than four times as likely as those with a college degree to report poor or fair health.[9] Adults living below the federal poverty level are more than four times as likely as those with incomes greater than 400 percent of the poverty level to indicate that they are in poor health.[10] Indeed, in the United States, the top half of male earners live 5.4 years longer than the bottom half.[11]

Recent healthcare reform measures extend insurance coverage to much, though not all, of the nation's uninsured population. Unfortunately, insurance reform alone cannot effectively reduce health disparities and improve the country's health status. The United States must transform its approach to health care: from a "sick care" system to one that focuses on prevention and wellness, from a hospital-based system to one in which primary care is central, and from fragmented service episodes to a true continuum of care across the life span.[12]

To accomplish this transformation, the current system in which medical care is isolated from other social services must be fundamentally changed. The nation must embrace a new approach that addresses the upstream social determinants that drive health status. Clinical care delivery is responsible for only about 10% of premature mortality and health status (Fig. 1). Other determinants of health are more powerful, including genetics, behaviors, and social circumstances, such as education, income, housing,

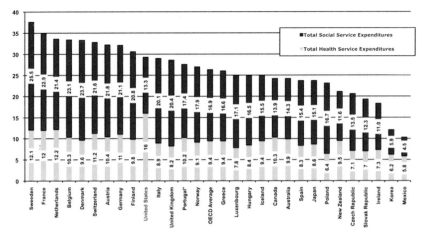

Figure 2. Health and social service expenditures by nation as percent of gross domestic product (GDP).

job security, transportation, safe neighborhoods, and access to nutritious foods.

Pomeroy argued that the concern that the country cannot afford the cost of addressing social determinants of health is ill-founded. The solution is not more money; instead, it is spending money on the right things at the right times. She stressed that at national and local levels, healthcare funds should be directed upstream toward social services to promote wellness and prevent disease, rather than downstream after disease develops.

As shown in Figure 2, the United States spends twice as much of its GDP on health care as other OECD countries, yet the total spent on health services plus social services (which can address the social determinants of health) is comparable to other countries.[13]

Successful examples of the upstream approach on the local level include healthcare systems that have instituted programs to provide housing for homeless patients who are ready for hospital discharge. Results demonstrate dramatic cost savings to the overall healthcare system, although not always to the organization bearing direct costs.[14]

Social determinants of health receive limited attention in current healthcare reform initiatives.[15] However, there is increasing awareness of the need to broaden the perspective to focus on health, rather than just health care. For example, the Association of Academic Health Centers recently adopted a resolution calling on its members to address social determinants of health.[16]

At the University of California (UC), Davis, Health System, model programs are in place that emphasize health disparities and social determinants in the mission areas of education, clinical care, and research, including:

- Education:
  - Rural-PRIME (programs in medical education) and San Joaquin Valley-PRIME prepare selected students to practice in rural communities, and TEACH-MS (Transforming Education and Community Health for Medical Students) programs train students and residents who are committed to careers in underserved urban areas.
  - Social determinants in the community are addressed through multiple outreach programs, including English and Spanish versions of mini-medical schools for seniors to support healthy aging.
- Clinical care:
  - UC Davis physicians work to overcome geographic disparities by providing consultations via telemedicine at more than 100 sites throughout California; they also helped to launch the California Telehealth Network, which links more than 800 sites across the state, including 12 e-health communities.
  - Electronic health records and personal health records for UC Davis patients include data on race, ethnicity, and primary language, and soon will include sexual orientation and gender identity.

- Research:
  - The UC Davis Center for Reducing Health Disparities leverages a multidisciplinary approach to studying the causes of health disparities and advances solutions that help address underlying social determinants.
  - The UC Davis Center for Population Health Improvement develops, applies, and disseminates knowledge about social determinants to improve health, health security, and health equity.
  - At the UC Davis Comprehensive Cancer Center, the Mothers' Wisdom Breast Health project has increased mammogram screening among American Indian and native-Alaskan women, and the National Center for Reducing Asian American Cancer Health Disparities has increased hepatitis testing and vaccination rates in at-risk communities.

There is a present need to address health disparities and focus on social determinants of health. Health professionals must be trained with the skills to excel in this new paradigm,[17] including working as interprofessional teams and in nontraditional venues of care, using innovative technologies, incorporating the science of systems care, advancing cultural competency, and embracing the social determinants of health. This new direction requires the collaborative support of academia, community leaders, and the public; health professionals cannot and should not move forward in isolation.

Pomeroy urged that all sectors of society must come together to create comprehensive and sustainable solutions. The causes of poor health and the solutions to improve health are multifaceted and interconnected. Integrated solutions that reflect a "health in all policies" approach are critical.[18] From business leaders to engineers, policy makers, social service experts, and more, the country must transcend traditional boundaries.[19] Collaboration opens doors to new perspectives, different questions and better answers. Together, Pomeroy declared, we will be able to make smarter policy decisions, create programs that focus on underlying social determinants of health, and improve health for all.

Culturally competent education

Improving quality and achieving equity through cross-cultural education

Joseph R. Betancourt (Massachusetts General Hospital, Harvard Medical School) discussed the importance of improving cross-cultural education among healthcare providers. The goal of cross-cultural education is to improve the ability of healthcare providers to communicate effectively and provide quality health care to patients from diverse sociocultural backgrounds. This field has emerged for two very practical reasons. First, the United States is becoming more diverse. As such, healthcare providers will increasingly see patients who may present their symptoms in different ways, who may have different thresholds for seeking care, and who may express different health beliefs that influence adherence. Second, research has demonstrated that sociocultural differences between patient and provider influence communication and clinical decision making, and are especially pertinent given the evidence that links provider–patient communication to patient satisfaction, adherence, and subsequently, health outcomes.[20] Thus, when sociocultural differences between patient and provider are not managed effectively in the medical encounter, patient dissatisfaction, poor adherence, poorer health outcomes, and lower quality care may result.[21]

The Institute of Medicine report *Unequal Treatment* highlighted the importance of cross-cultural communication—and recommended cross-cultural medical education—as a means of eliminating racial/ethnic disparities in health care.[20] Disparities can result from poor communication between providers and minority patients, minority patient mistrust, and stereotyping of minority patients by providers. In fact, a recent meta-analysis conducted by the Agency for Healthcare Research and Quality, consisting of a systemic review of 91 articles that measured the impact of cultural competence training on the quality of care provided to minority patients, found that this training yielded improvement in provider knowledge, attitudes, and skills in this area, as well as improvements in patient satisfaction.[21]

In order to assess how prepared health professional trainees feel they are to care for diverse populations, a national survey was conducted with more than 2000 residents in their last year of training in seven specialties (medicine, surgery, pediatrics, obstetrics and gynecolgy, psychiatry, emergency medicine, and family medicine).[22] Nearly all residents indicated that it was important to consider the patient's culture when providing care, and many residents indicated that cross-cultural issues often resulted in negative consequences for clinical care, including longer office visits, patient noncompliance, delays in obtaining consent, unnecessary tests, and lower quality of care. Interestingly, approximately one in five residents indicated that they possessed low skills in this area. When compared to residents who had reported receiving a lot of cross-cultural education, those who reported receiving little or none were 8–20 times more likely to report low skill levels in a variety of key areas essential to the care of diverse populations.

Betancourt summarized by stressing that culture matters in clinical care, being inattentive to culture has significant quality implications, and those who receive cross-cultural education feel more prepared to handle these challenges. Despite this, limited time in the clinical encounter remains the largest barrier to implementation of cross-cultural skills when present. Betancourt stressed that his experience teaching cross-cultural communication to medical students, residents, and practicing clinicians taught him that providers want to do the right thing; however, they do not want to be lectured or told that they are culturally incompetent and need to be fixed. Furthermore, medical students and physicians often want just the basic facts about culture and often view cultural competence as something that increases visit time rather than a skill set, and as a soft science without an evidence base. Therefore, Betancourt argued, cross-cultural competence needs to be framed as a skill set—similar to a review of systems, or checklist—that can help providers manage challenging cross-cultural cases. It must be seen as practical, actionable, and time efficient; it needs to be taught in a case-based fashion that creates clinical challenges; it must be linked to evidence-based guidelines and the peer-reviewed literature; and it must leave learners with a concrete set of tools and skills.

To meet this need, Betancourt and colleagues developed a portfolio of cross-cultural communication e-learning programs entitled Quality Interactions.[23] Since 2003, approximately 125,000 healthcare professionals have completed one of these programs (including about 1000 doctors at Massachusetts General Hospital in the span of three months in 2009). Built on 15 years of research and educational experiences, Quality Interactions teaches a patient-based approach to cross-cultural care.[24] This approach is based on the premise that every individual patient presents a sociocultural perspective that must be explored and managed on an as-needed basis through a set of questions, inquiries, and negotiation strategies. Betancourt argued that e-learning is a powerful and effective tool for cross-cultural education at all levels. It allows for extensive training of a large group of learners in a short amount of time with a set of uniform skills. The Quality Interactions program has demonstrated that e-learning needs to be case based, interactive, and capable of creating teachable moments; it must provide personalized feedback; and it should be longitudinal (with boosters) and present the option for blended learning (classroom teaching as a supplement). E-learning programs need to be realistic, easy to maneuver, linked to evidence-based guidelines and the peer-reviewed literature, and to provide skills to more effectively communicate with all patients.

In conclusion, the field of cross-cultural education is growing rapidly. We must be prepared to meet the needs of an increasingly diverse population by creating a skilled healthcare workforce that can deliver effective, high-quality care. Although healthcare providers may be somewhat resistant to cross-cultural education, this can be overcome by messaging the importance of this work and its link to quality of care. In the setting of faculty, time, and financial resources, e-learning provides an excellent mechanism for extensive, high-quality cross-cultural education for medical students, residents, and practicing clinicians.

Educating students to deliver comprehensive LGBT care

Lesbian, gay, bisexual, and transgender (LGBT) populations, which represent approximately 7% of the United States' population, are increasingly recognized as groups with documented health and healthcare disparities. Mitchell R. Lunn, (Brigham and Women's Hospital, Harvard Medical School) discussed inequities, including barriers to accessing health care and increased risk of certain chronic diseases, as well as unique physical and mental health challenges related to sexual orientation and gender identity.[25] In 2007, the Association of American Medical Colleges (AAMC) recommended that medical schools "ensure that students master the knowledge, attitudes, and skills necessary to provide excellent, comprehensive care for [LGBT] patients."[26] Despite these recommendations, most LGBT patients believe providers are not prepared to care for them.[27] Prior LGBT-related medical education studies, while limited in scope, highlighted the evolution of the medical community and society; gay and lesbian studies were initially conducted within psychiatry and later within family medicine. These studies highlighted the importance of gay and lesbian student groups in medical education and showed that increased clinical exposure to LGBT patients correlated with greater LGBT health knowledge and skills as well as more positive attitudes toward LGBT patients.

Lunn presented the findings from a 2011 national survey study of deans of medical education, with a 75% response rate, that provided a contemporary, comprehensive estimate of LGBT-related medical education.[28] The median number of hours dedicated to teaching LGBT-related content in the required medical school curriculum was five. More than one-third of respondents reported zero hours during the clinical years. While most institutions cover sexual orientation, gender identity, and HIV, the least-taught topics included those related to primary care (e.g., substance abuse, chronic disease risk, body image) and transgender people (e.g., transitioning, sex reassignment surgery). Overall, most deans had a negative opinion about the amount and quality of the coverage of LGBT content at their institutions. In order for improvement, they noted a need for more LGBT health curricular materials and faculty willing to teach the content.

Given that the teaching of these topics is often limited to a select few interested faculty members, developing new curricula for LGBT health can be a daunting task (Table 1). Lunn emphasized the importance of using lectures, cases, simulations, and observed structured clinical encounters (OSCEs) designed by others until LGBT health competencies broadly influence curricular development. Several national LGBT health-focused organizations—including the Gay and Lesbian Medical Association (www.glma.org) and the Fen-

Table 1. Suggestions for improving LGBT education by improving educational environment

Increase awareness
- LGBT student and/or employee group
- Out list of LGBT-identified faculty, students, and staff
- Dedicated LGBT center
- LGBT-specific admission materials
- Faculty governing body support of marriage equality
- Anti-discrimination policies
- Rainbow sticker campaign

Increase didactic and experiential learning
- Use existing curricular resources on MedEdPORTAL and the Fenway Institute
- Gay and Lesbian Medical Association "Recommendations for LGBT Equity and Inclusion in Health Professions Education"
- Simulation cases
- Clinical case discussions with LGBT patients
- Standardized patient cases with LGBT patients
- LGBT-focused community health centers

Increase personnel development
- Institutional mission statement including supporting communities and their people
- Institutional commitment to diversity
- Recruit and retain LGBT faculty and students
- Develop promotion and tenure pathways for LGBT health disparity and population work
- Funding for LGBT health disparity and population work
- Coursework for students interested in LGBT health disparity and population work

way Institute (www.thefenwayinstitute.org)—have developed curricular materials on LGBT health. Additional national repositories, like the AAMC's MedEdPORTAL (www.mededportal.org), allow educators to submit their curricula for peer review and facilitate the exchange of teaching resources throughout undergraduate and graduate medical education.

In addition to didactic and experiential learning, improving LGBT medical education requires increased awareness and personnel development (Table 1).[29] Developing the students and faculty who will improve LGBT health disparities by research and teaching requires an institutional commitment to diversity. Targeted student and faculty recruitment, mentorship programs, and defined pathways for faculty promotion may help establish a culture that accepts health disparities research as scholarship while creating a welcoming learning environment. LGBT-specific efforts (e.g., Out List, LGBT center, rainbow sticker campaign, support for civil marriage equality) at institutions across the country not only promote the visibility of LGBT people but also normalize sexual orientation and gender identity disclosure. By creating diverse educational environments with persons needed to drive change, LGBT-related medical education will improve and, as a result, so will the care of LGBT people.

Changing curricula

Addressing health disparities through molecular epidemiology

At its current rate of growth, the racial and ethnic composition of the U.S. population will continue a dynamic shift toward increasing diversity. Over the past two decades, while significant advances in medical, diagnostic, and therapeutic innovations have been made in key areas, such as infectious disease, cardiovascular health, and oncology, significant inequities remain in health outcomes according to factors such as race and ethnicity, sexual and gender orientation, income, and education. The social environment in which individuals live, as well as their lifestyles and behaviors, can influence the incidence of illness in populations.[30] At no time has this been more evident than in the current climate of recognized health disparities, lack of health insurance, and economic instability. Having an understanding of these inequities and being equipped to address modifiable determinants of health disparities provides the physician-in-training with the necessary foundation to effectively practice medicine in the 21st century and to address important public health needs. Thus, the training of the modern physician must weave the development of traditional skills such as communication and physical examination with the cost-effective use of novel biomolecular information in complex systems to deliver culturally competent care. The promise of this training is the development of individuals who are at the forefront of using translational medicine to address disparities in health.

Fritz Francois (New York University School of Medicine (NYUSoM)) discussed the introduction, in 2010, of a patient-centered curriculum for the 21st century (C21 at NYUSoM) with exemplars

of disease or pillars that span from the preclinical period (shortened to 18 months) to the clinical years. Colorectal cancer serves as an example of a cancer biology pillar that provides students with the opportunity to connect information from the classroom, the bench, and the bedside, as they learn anatomy, histology, pathology, and epidemiology, among other subjects, while considering both modifiable as well as nonmodifiable determinants of disease development. Through case scenarios, students are challenged with specific tasks such as counseling a patient about colon cancer screening where both patient[31] and physician barriers[32] need to be considered. The student engages in team-based decision-making exercises that incorporate issues in population health and health disparities, while using biomolecular evidence to better understand possible causes for the clinical scenario as well as to consider the costs and benefits of possible solutions. The students are provided with new educational opportunities as well as student-centered curricular pathways, such as concentrations that allow focus in a specific area as they explore the determinants of disease.

One such curricular pathway, implemented at NYUSoM in the fall of 2012, is the Health Disparities Concentration (HDC). It was created to provide a forum for learning about determinants of health disparities and for developing skills that help promote health equity through clinical practice and scholarly work. Consistent with the 2011 U.S. Centers for Disease Control and Prevention (CDC) Health Disparities and Inequalities Report (http://www.cdc.gov/minorityhealth/reports/CHDIR11/ExecutiveSummary.pdf), the didactic clinical exposure and scholarly work components of the concentration are organized along six central categories: social determinants of health, environmental hazards, healthcare access and preventive health services, mortality, morbidity, and behavioral risk factors. The central categories are aligned with the C21 pillars to allow for the spiraling of content areas to which the students have already been exposed. The HDC allows for the exploration of multifaceted aspects of the central categories, including policy development, legal implications, economic considerations, and community engagement. The specific goals for the HDC participant are (1) to understand and be able to describe determinants of health disparities, (2) to gain exposure to and participate in healthcare delivery in an underserved area, and (3) to examine and/or apply a strategy to achieve health equity through scholarly work. This immersion experience provides the student with the opportunity to make use of biomolecular information to understand and to possibly address the epidemiology of disparities in health observed in the community.

Promoting diversity

From fairness to excellence: making diversity matter

Marc A. Nivet (Association of American Medical Colleges) characterized the present as a major turning point for health care in America. The many systemic changes tied up with the Accountable Care Act and ongoing efforts to stabilize costs while continuing to improve the quality of care to a broader population mean inevitable upheaval for healthcare institutions. As the educators of future physicians and providers of advanced care and research, the academic medicine community has a critical role to play in achieving meaningful change during this turbulent process.

Disparities in health status, access to care, and quality of care received remain prevalent across our society. Nivet argued that excellence in health care for all should be a goal at the forefront of our minds as we reform for the future. And, while health disparities are largely attributable to the upstream social determents of health,[33,34] the field of medicine has a crucial role to play in achieving health equity; after all, the mission and purpose of the profession is to meet the health needs of society.

One important aspect in addressing health disparities is developing the appropriate workforce to understand and address the needs of an increasingly diverse patient population.[35] While longstanding efforts to address these issues focused almost entirely on increasing the numbers of racial and ethnic minorities in the health professions, efforts should now recognize how diversity contributes across the board to helping institutions meet their targets of excellence. This means coupling efforts to boost compositional diversity with a transformation of organizational culture to increase engagement and inclusion. Educational environments and institutional climates need to be explicitly structured to maximize the benefits of diversity, as decades of experience have shown that the passive presence of

Figure 3. A model for the embrace of diversity and inclusion as a significant component of the mission to achieve health equity.

diversity within organizations, especially below a critical mass, is not sufficient to realize its deepest potential value.[36]

At many academic medical centers, diversity efforts remain detached from the core mission and run parallel to key strategic organizational processes.[37] This siloed structure reflects a model of diversity as a series of problem to solve or wrongs to right. The necessary shift is one from fairness to excellence, embracing diversity and inclusion as an essential component of meeting the mission and delivering top notch care (Fig. 3).

Despite decades of work to promote diversity within academic medicine, relatively few evaluations have been conducted to determine the most effective specific interventions. Because of the associations with charged issues and relegation to the organizational periphery, the value of diversity and inclusion efforts has often been accepted at face value, with interventions disconnected from a rigorous evidence base. Viewed in the context of efforts to bend the cost curve—improving health care for all, reducing health disparities, and reversing the upward cost spiral—optimizing the impact of our investments in diversity and linking them to mission excellence is critical.

Promoting diversity in healthcare professionals

There are many routes to eliminating health disparities: one of them is through education and

Figure 4. The healthcare continuum from research to translation to implementation. Training a new workforce in translational research may be key to addressing disparities in health care.

career development of a new generation of translational researchers in the health workforce.[39,40] We aim to prepare this workforce by conveying research skills and knowledge through translational research in health disparities to implement knowledge acquired in a multidisciplinary teamwork environment with the goal of achieving health for all. This concept was envisioned by Estela S. Estapé (University of Puerto Rico) as a continuum from research to translation to implementation, toward better health for all. (Fig. 4). Estapé and Mekbib Gemeda (New York University) discussed strategies and educational programs that are actively working to attract underrepresented minorities to medical and scientific professions.

Gemeda highlighted the very evident underrepresentation of minorities in academic medicine as faculty or researchers. Traditionally, minorities who do receive medical education have been encouraged to work in underserved communities, rather than to pursue academic research. This trend continues to the present. While he lauded these students' commitment to community service, he argued that such work was not mutually exclusive with academic medicine. The key to getting this message across, Gemeda said, was curriculum innovation: to integrate health disparities education into medical school curricula; to focus on the social determinants of health; to understand interventions across social, behavioral, clinical, and policy paradigms; and to engage students in community-based research and education, so that students can experience how academic institutions can function in cooperation with communities. These efforts, he argued, were the likeliest means of truly encouraging diversity within academic medicine.

Estapé described a new postdoctoral masters of science program in clinical and translational research that she is spearheading at the University of Puerto Rico, which was developed as a joint program between two academic units: the School of Health Professions and the School of Medicine.[41] This program continues to expand and strengthen its reach, curriculum, and leadership, through the support of NIH, to diversify the workforce and to transfer knowledge into practice for better health outcomes while making health care a more efficient and effective process and reducing costs.

With regard to medical education and its role in advancing translational research, Estapé addressed several questions from her perspective as a leader in developing translational research education collaborations and partnerships:

When to start? At the predoctoral level, during the first two years of medical education, is the time to build the basic core values upon which to develop a research career. During these years, medical students need to acquire and demonstrate basic core values needed to excel in health service and to engage in eliminating health disparities: integrity, compassion, trust, hope, humility, and respect.

What to offer? While becoming a physician, the student should have opportunities in the curriculum that include core knowledge about health disparities and experiences that will facilitate active participation in community service, as well as activities that are planned to develop critical thinking skills. During the last two years of the medical curriculum, students who are interested in pursuing research careers should be provided with opportunities to work in a team both as a member and leader, to be a peer mentor or advisor to others, and to have active participation in mentored research activities.

Who advances? There are at least three criteria that will help the medical graduate in pursuing the ideal of becoming a translational researcher: self-motivation, passion to make a difference in the health of others, and, of utmost importance, strong mentors and support. Estapé emphasized that research career development is more productive and efficient after the medical graduate has determined an area of interest (knows what he/she wants to become) and what type of research is more suitable for her work. From her experience, Estapé believes that one should "only look to the past (evidence, knowledge) to create the future (hypothesis, research)." That evidence is expected to be a guide on the way to effective and efficient translational research.

When to advance? Although the foundation for a career as a researcher in translational research is best received during medical studies at the predoctoral level, formal studies for becoming a clinical and translational researcher are most productive at the postdoctoral level.

What skills and competencies? NIH has defined the core competencies needed to be part of the next generation of clinical and translational researchers. These 14 core thematic areas and 101 competencies define the basic knowledge, skills, and attributes that a master's level candidate should attain.[41] The level at which the core thematic areas are covered in a curriculum will determine the competencies expected from the graduate and her success as an independent clinical and translational researcher.

Therefore, an effective way to make a significant contribution toward the elimination of health disparities is through the development of innovative multidisciplinary research degree programs with mentoring, collaborations, and formation of research teams as the essence for advancing clinical practice and outcomes. Some of the challenges to be met in advancing implementation are to break down institutional barriers, foster team science, facilitate public–private partnerships, promote collaborations and partnerships, network, and live with a positive attitude.

Student presentations

A panel of medical students made a presentation at the conference, giving attendees the opportunity to hear the perspectives of students who are participating in and promoting health disparities education and research at their respective institutions and beyond.

The students brought a wide variety of personal experiences. Temitope Awosogba, (third-year MD student at the Mount Sinai School of Medicine) has explored her interest in health disparities research by conducting primary epidemiologic research in Mozambique and serving as an administrator, clinician, and educator at her school's free student-run clinic for the uninsured. Howa Yeung, (fourth-year MD student at NYUSoM), collaborated with several other students and the director of the Health, Medicine and Society course at the City University of New York (CUNY) Sophie Davis School of Biomedical Education to develop a novel, problem-based learning curriculum for the course. This innovative case-based curriculum used a series of workshops to demonstrate the methods fundamental to community health and disparities research. F. Garrett Conyers (second-year MD–PhD student at the Harvard Medical School) has worked extensively with faculty and administrators to incorporate health disparities into the curriculum. Sabrina Gard (fourth-year MD student at NYUSoM) earned her masters of public health degree in order to learn how to apply the principles of primary care public health to alleviating health disparities.

The students all agreed that the fundamental significance of making health disparities education and research a regular part of the curriculum lies in the manner in which those educational experiences are translated to patient care. First, these educational experiences allow students to develop cultural competence. Cultural competence in the healthcare system has been described by Betancourt *et al.* as acknowledging and incorporating the importance of culture, expansion of cultural knowledge, and adaptation of services to meet culturally unique needs.[42] Students are taught that in order to address racial/ethnic health disparities in the United States, the health system, including the healthcare professionals who operate within it, must be more culturally competent. Understanding, for example, why an individual or population doesn't access care until late stages of disease, or why diet choices are made, does more for addressing health inequity than more paternalistic medicine. Health disparities research in medical education exposes students to the importance of thinking about health, not only as a function of individual decisions but also as a function of community, access, the built environment, and other social determinants of health and disease.

The question of whether health disparities education and/or research should be required in medical education was a point of some contention for the students. All four students believed that health disparities education should be required in the medical school curriculum, stating that the lack of student exposure to health disparities results in a lack of appreciation for the topic by trainees. Conyers felt strongly that health disparities resulted from decades of structural violence against marginalized communities, and that trainees must be required to study health disparities in order to unmask the institutional causes of disparities. Awosogba agreed that health disparities should be a part of medical school education due to the fundamental injustice inherent in the existence of health disparities. She argued that because health disparities may not result solely from biological differences between groups of people, differences in the social, economic, political, and physical environment must be better understood. As frontline members for delivering healthcare, physicians are in a unique and powerful position to influence the impact of these social injustices. On the other hand, given the time and dedication required in conducting and completing substantial research projects, Gard and Yeung argued that health disparities research infrastructure and support should be tailored to medical students expressing high levels of interest in the field and not be required for all. The students also discussed the manner in which health disparities have already been incorporated into their medical education through lectures, small group discussions, and OSCEs, but have not been addressed as an issue of injustice or active violence against marginalized communities.

While the students agreed on the significance of incorporating health disparities education into medical school curricula, they also elucidated a number of barriers to the implementation of these programs. Awosogba and Yeung discussed the current lack of buy-in from medical schools: stakeholders must be convinced that health disparities are a necessary part of the curriculum and that health disparities education will produce better, more competent clinicians before structured education or research programs in health disparities can be initiated and integrated into medical education. Support for health disparities education and research opportunities also requires financial support, faculty mentorship, administrative support, and structured research programs. The students were collectively optimistic about the steps that are being taken, largely by their peers in medical schools across the country, to incorporate health disparities into medical school curricula and research opportunities.

Acknowledgements

The conference "Prioritizing Health Disparities in Medical Education to Improve Care" was presented by the Josiah Macy Jr. Foundation, the Associated Medical Schools of New York, the New York University School of Medicine, and the New York Academy of Sciences.

E. Estapé acknowledges the NIH/NIMHD for their support to continue building research infrastructure and capacity in Puerto Rico: Endowment Program "Hispanics in Research Capability: School of Health Professions & School of Medicine Partnership (HiREC)" S21MD001830, and the Hispanic Clinical and Translational Research Education and Career Development (HCTRECD) program R25MD007607.

Conflicts of interest

The authors declare no conflicts of interest.

References

1. National Center for Health Statistics. 2011. *Health, United States, 2011: With Special Feature On Socioeconomic Status and Health.* Hyattsville.
2. Smedley, B.D., A.Y. Stith & A.R. Nelson, Eds. 2003. *Unequal Treatment: Confronting Racial and Ethnic Disparities in Health Care.* National Academies Press. Washington, D.C.
3. Geppert, C.M.A., C.L. Arndell, A. Clithero, *et al.* 2011. Reuniting public health and medicine: The University of New Mexico School of Medicine Public Health Certificate. *Am. J. Prev. Med.* **41:** S214–S219.

[This article was corrected on 23 May 2013 after original online publication. Additional Acknowledgements were added.]

4. Pacheco, M., D. Weiss, K. Vallant, *et al.* 2005. The impact on rural New Mexico of a family medicine residency. *Acad. Med.* **80:** 739–744.

5. Kaufman, A., W. Powell, C. Alfero, *et al.* 2010. Health extension in New Mexico: an academic health center and the social determinants of disease. *Ann. Fam. Med.* **8:** 73–81.

6. Johnson, D., P. Saavedra, E. Sun, *et al.* 2011. Community Health Workers and Medicaid Managed Care in New Mexico. *J. Comm. Health.* **37:** 563–571.

7. Organisation for Economic Co-operation and Development. 2012. *OECD Health Data 2012- Frequently Requested Data.* February 13, 2012. Cited March 12, 2013. http://www.oecd.org/els/health-systems/OECDHealthData2012Freque ntlyRequestedData_Updated_October.xls

8. Drum, C.E., M.R. McClain, W. Horner-Johnson & G. Taitano. 2011. *Health Disparities Chart Book on Disability and Racial and Ethnic Status in the U.S.* Institute on Disability, University of New Hampshire. Durham.

9. Adams, P.F., M.E. Martinez, J.L. Vickerie & W.K. Kirzinger. 2010. *Summary Health Statistics for the U.S. Population: National Health Interview Survey.* Washington D.C. National Center for Health Statistics.

10. Hampton, T. 2008. Group seeks to improve nonmedical aspects of health in the United States. *J. Am. Med. Assoc.* **299:** 1761–1762.

11. Waldron, H. 2007. Trends in Mortality Differentials and Life Expectancy for Male Social Security–Covered Workers by Socioeconomic Status. *Social Security Administration Bulletin* 67.

12. Asch, D.A. & K.G. Volpp. 2012. What business are we in? The emergence of health as the business of health care. *New Engl. J. Med.* **367:** 888–889.

13. Bradley, H.M., E.R. Benjamin, J. Herrin & E. Brian. 2011. Health and social services expenditures: associations with health outcomes. *BMJ Qual. Saf.* **20:** 826–883.

14. Kuehn, B.M. 2012. Supportive housing cuts costs of caring for the chronically homeless. *JAMA* **308:** 17–19.

15. Bruno, K. & J.E. Grigsby. 2010. *The Role of Academic Health Centers in Addressing the Social Determinants of Health.* Atlanta. The Blue Ridge Academic Health Group.

16. Association of Academic Health Centers. 2012. AAHC Board Endorses Commitment to Address the Social Determinants of Health. April 24, 2012. Cited March 12, 2013. http://www.aahcdc.org/Policy/PressReleases/PRView/ArticleId/106/AA HC-Board-Endorses-Commitment-to-Address-the-Social-Determinants-of-Health.aspx.

17. Barnes, K.A., J.C. Kroening-Roche & B.W. Comfort. 2012. The developing vision of primary care. *New Engl. J. Med.* **367:** 891–893.

18. The Aspen Institute. 2009. *Health Stewardship: The Responsible Path to a Healthier Nation.* Washington, D.C. The Aspen Institute.

19. Bawa, K.S., G. Balachander & P. Raven. 2008. A case for new institutions. *Science* **319:** 136.

20. Smedley, B.D., A.Y. Stith & A.R. Nelson, Eds. 2003. *Unequal Treatment: Confronting racial and ethnic disparities in health care.* Washington, D.C. National Academy Press.

21. Beach, M.C., L.A. Cooper. K.A. Robinson, *et al.* 2004. Strategies for improving minority healthcare quality. *Evid. Rep. Technol.y Assess.* (Summ) **90:** 1–8

22. Weissman, J.S., J. Betancourt, E.G. Campbell, *et al.* 2005. Resident physicians' preparedness to provide cross-cultural care. *J. Am. Med. Assoc.* **294:** 1058–1067.

23. Quality Interactions. 2013. Available at: http://www.quality interactions.org/.

24. Carrillo, J.E., A.R. Green & J.R. Betancourt. 1999. Cross-cultural primary care: A patient-based approach. *Ann. Intern. Med.* **130:** 829–834.

25. Gay and Lesbian Medical Association and LGBT Health Experts. 2010. *Healthy People 2010 Companion Document for Lesbian, Gay, Bisexual, and Transgender (LGBT) Health.* San Francisco, CA: Gay and Lesbian Medical Association; April, 2010. Cited March 12, 2013. http://www.glma.org/_data/n_0001/resources/live/HealthyCompanionDoc3.pdf

26. Joint AAMC-GSA and AAMC-OSR Recommendations Regarding Institutional Programs and Educational Activities to Address the Needs of Gay, Lesbian, Bisexual and Transgender (GLBT) Students and Patients. 2007. Association of American Medical Colleges. March 1, 2007. Cited March 12, 2013. https://www.aamc.org/download/157460/data/institutional _programs_and_educational_activities_to_address_th.pdf

27. When Health Care Isn't Caring: Lambda Legal's Survey of Discrimination Against LGBT People and People with HIV. 2010. Lambda Legal. http://data.lambdalegal.org/publications/downloads/whcic-report_when-health-care-isnt-caring.pdf

28. Obedin-Maliver, J.E., E.S. Goldsmith, L. Stewart, *et al.* 2011. Lesbian, Gay, Bisexual, and Transgender-Related Content in Undergraduate Medical Education. *J. Am. Med. Assoc.* **306:** 971–977.

29. Lunn, M.R. and J.P. Sanchez. 2011. Prioritizing Health Disparities in Medical Education to Improve Care. *Acad Med.* **86:** 1343.

30. Institute of Medicine. The future of public health. 1988. Washington, D.C. National Academy Press.

31. Francois, F., G. Elysee, S. Shah & F. Gany. 2009. Colon cancer knowledge and attitudes in an immigrant Haitian community. *J. Immigr Minor. Health.* **11:** 319–325.

32. White, P.M., M. Sahu, M.A. Poles & F. Francois. 2012. Colorectal cancer screening of high-risk populations: A national survey of physicians. *BMC Res. Notes.* **5:** 64.

33. Williams, D.R., M.V. Costa, A.O. Odunlami & S.A. Mohammed. 2008. Moving upstream: how interventions that address the social determinants of health can improve health and reduce disparities. *J Public Health Manag. Pract.* **14:** S8–S17.

34. Satcher, David. 2010. Include a Social Determinants of Health Approach to Reduce Health Inequities. *Public Health Rep.* **125:** 6–7.

35. Health Resources and Services Administration, Bureau of Health Professions. The Rationale for Diversity in the Health Professions: A Review of the Evidence. 2006. Rockville, Maryland. U.S. Dept. of Health and Human Services.

36. Milem, J.F., M.J. Chang & A.L. Antonio. 2006. *Making Diversity Work on Campus: A Research Based Perspective.* Washington, DC: Association of American Colleges and Universities.

37. Nivet, M.A. 2011. Commentary: diversity 3.0: a necessary systems upgrade. *Acad. Med.* **86:** 1487–1489.

38. Selker, H.P. 2010. Beyond translational research from T1 to T4: beyond "separate but equal" to integration (Ti). *Clin. Transl. Sci.* **3:** 270–271.

39. Dankwa-Mullan I., K.B. Rhee, D.M. Stoff, *et al.* 2010. Moving toward paradigm-shifting research in Health disparities through translation, transformation and transdisciplinary approaches. *Am. J. Public Health.* **Suppl 1:** S19–24.

40. Estape, E., B. Segarra, A. Baez, *et al.* 2011. Shaping a New Generation of Hispanic Clinical and Translational Researchers Addressing Minority Health and Health Disparities. *Puerto Rico Health Sci. J.* **30:** 167–175.

41. Clinical and Translational Sciences Awards. 2009. Core Competencies in Clinical and Translational Research. July 14, 2009. Cited March 12, 2013. https://www.ctsacentral. org/education_and_career_development/core-competencies-clinical-and-translational-research

42. Betancourt, J.R., R.R. Green, J.E. Carrillo, *et al.* 2003. Defining Cultural Competence: A Practical Framework for Addressing Racial/Ethnic Disparities in Health and Health Care." *Public Health Rep.* **118:** 293–302.

Ann. N.Y. Acad. Sci. ISSN 0077-8923

The paradox of overnutrition in aging and cognition

Roger A. Fielding,[1] John Gunstad,[2] Deborah R. Gustafson,[3,4] Steven B. Heymsfield,[5] John G. Kral,[6] Lenore J. Launer,[7] Josef Penninger,[8] David I. W. Phillips,[9] and Nikolaos Scarmeas[10,11]

[1]Nutrition, Exercise Physiology, and Sarcopenia Laboratory, Tufts University, Boston, Massachusetts. [2]Department of Psychology, Kent State University, Kent, Ohio. [3]Department of Neurology, SUNY Downstate Medical Center, Brooklyn, New York. [4]Neuropsychiatric Epidemiology Research Unit, University of Gothenburg, Gothenburg, Sweden. [5]Pennington Biomedical Research Center, Louisiana State University, Baton Rouge, Louisiana. [6]Departments of Surgery and Medicine, SUNY Downstate Medical Center, Brooklyn, New York. [7]Laboratory of Epidemiology, Demography, and Biometry, Intramural Research Program, National Institute on Aging, Bethesda, Maryland. [8]Institute of Molecular Biotechnology, Austrian Academy of Sciences, Vienna, Austria. [9]Lifecourse Epidemiology Unit, University of Southampton, Southampton, United Kingdom. [10]Department of Neurology, Columbia University College of Physicians and Surgeons, New York, New York. [11]University of Athens, Athens, Greece

Address for correspondence: John G. Kral, Department of Surgery, SUNY Downstate Medical Center, 450 Clarkson Avenue, Box 40, Brooklyn, NY 11203-2098. jkral@downstate.edu

Populations of many countries are becoming increasingly overweight and obese, driven largely by excessive calorie intake and reduced physical activity; greater body mass is accompanied by epidemic levels of comorbid metabolic diseases. At the same time, individuals are living longer. The combination of aging and the increased prevalence of metabolic disease is associated with increases in aging-related comorbid diseases such as Alzheimer's disease, cerebrovascular dementia, and sarcopenia. Here, correlative and causal links between diseases of overnutrition and diseases of aging and cognition are explored.

Keywords: overnutrition; obesity; cognitive decline; Alzheimer's disease; sarcopenia

Introduction

Over one billion people are estimated to be overweight, placing them at risk for diabetes, cardiovascular disease, and cancer. Type 2 diabetes (T2D) and related metabolic diseases have now become epidemic in resource-poor countries that have undergone rapid westernization and where the consumption of energy-rich diets, coupled with reduced physical activity, has led to an increase in the incidence of obesity. The combination of T2D, related metabolic diseases, and obesity, together with their greatly increased risk of cardiovascular disease, pose a major challenge to health services and the global economy.

Indicators of overnutrition are linked to both clinical dementia and Alzheimer's disease (AD), as well as to overall mortality. AD and other dementias are the fifth leading cause of death in people aged ≥65 years in the United States, according to the Centers for Disease Control and Prevention. One in three seniors currently dies from dementia. The risk of death in AD has steadily been increasing during the last three decades; it rose 39% from 2000 to 2010. Epidemiological studies in middle and later life have found that higher levels of body mass index (BMI) and waist circumference or waist-to-hip ratio (WHR), indicators of central adiposity, are associated with increased risk for dementia, whereas decline in body weight or BMI, and subsequent underweight, in the years preceding and at the time of a dementia diagnosis also relate to increased risk of dementia. The role of excess regional adipose tissue during different periods of life in relation to later and concurrent risk for cognitive decline and dementia as well as to overall aging is not understood. It has negative metabolic effects on peripheral and cerebral vasculature: adipose tissue (adipokines) and related hormones affect brain nuclei important for cognition, energy metabolism, and influence genetic susceptibility.

doi: 10.1111/nyas.12138

On December 4, 2012, the Sackler Institute for Nutrition Science and the New York Academy of Sciences hosted the conference "The Paradox of Overnutrition in Aging and Cognition" to present cutting-edge research that links and investigates overnutrition, aging, and cognition. Various strategies, including the use of imaging and metabolic markers and the identification of predictors of aging-related phenomena such as physical impairment need to be employed to understand and affect the linkages between overnutrition and the aging process, especially its cognitive aspects. The meeting, which included students and scientists working on aging and nutrition research and its application to preventive and curative medicine, presented and discussed clinical and epidemiological aspects of the relationships between overnutrition, aging, and cognitive performance. The conference was structured as three sessions: (1) an exploration of genetics and body composition with regard to overnutrition and pathology, chaired by John G. Kral (SUNY Downstate Medical Center); (2) an examination of the links between cognition, aging, and diabetes, chaired by Deborah R. Gustafson (SUNY Downstate Medical Center, and the University of Gothenburg); and (3) a discussion of ways to measure and modify the risks of overnutrition, also chaired by Kral. The meeting concluded with a panel discussion including most of the participants and facilitated by Kral.

Genetics and body composition

Systems biology approaches to overnutrition and aging

Nearly all essential metabolic regulators (e.g., insulin and mTOR signaling) exert conserved functions across different phyla. Similar to mammals, *Drosophila melanogaster* and *Caenorhabditis elegans* utilize a variety of evolutionarily conserved metabolic networks to control the equilibrium of energy and nutritional states. In *C. elegans*, an RNAi feeding model has been used to classify novel regulators of fat storage.[1] Recently, a genome-wide library of transgenic-RNAi lines of *D. melanogaster* has been developed in which the expression of any gene can be temporally and spatially controlled.[2]

Josef Penninger (Austrian Academy of Sciences) described a genome-wide RNAi screen designed to genetically dissect essential pathways of adiposity in adult *D. melanogaster*. Using tissue-specific (e.g.,

muscle, oenocyte, fat body, and neuronal) gene expression technology (i.e., tissue-specific GAL4 promoters), Penninger and colleagues classified candidate obesity genes according to function.[3] Among multiple known and novel candidate pathways, Hedgehog signaling was the most prominent fat body–specific obesity pathway. To translate these findings to mammals, tissue-specific mutant mice were generated to have a constitutively active Hedgehog signaling pathway in fat cells; the mice (aP2-*Sufu*) displayed near total loss of white, but not brown, adipose tissue. Mechanistic studies by Penninger's group revealed that activation of Hedgehog signaling blocked differentiation of white adipocytes through dysregulation of early adipogenic factors. Activation of the Hedgehog pathway did not, however, affect the differentiation of brown adipocytes in cell culture systems or in mutant mice.[3] This work identified a novel role for Hedgehog signaling in the determination of white/brown adipocytes. Moreover, the genetic screen provided a proof-of-principle that results obtained via unbiased *in vivo* RNAi-based scanning of the *D. melanogaster* genome can be used to explore adipocyte cell fate in mammals.

On the basis of these initial findings, the Hedgehog signaling pathway was also evaluated in mature, differentiated adipocytes and *in vivo* in mice.[4] These experiments identified a novel Ca^{2+}/G protein–coupled Hedgehog–Smo–AMPK axis that triggers a rapid Warburg-like aerobic glycolysis state. Small molecule Smo modulators uncoupled the Hedgehog–Smo–AMPK axis from the classical Hedgehog activation cascade; also, induction of the novel Smo–AMPK axis resulted in rapid glucose clearance in muscle and brown adipose tissue *in vivo*, and activation of the pathway bypassed the requirement for insulin in type 1 diabetic mice. These data demonstrate that Hedgehog signaling can rewire cellular metabolism in the context of glucose dysregulation, and thus may provide a unique therapeutic avenue for obesity and type 1 diabetes.[4]

With the advance of new sequencing technology, thousands of candidate genes have been identified as being involved in the basic control of physiology and disease pathogenesis. In *D. melanogaster* and *C. elegans*, large-scale screens have been performed examining pathologies related to adiposity. A key goal of functional disease genomics is to more rapidly translate the wealth of genetic associations found in model organisms to studies in mammals, especially as the data from experimental systems have used methodologies such as RNAi screens or time-consuming breeding of genetically altered mice.

Penninger and colleagues have introduced a novel technology with the potential to revolutionize functional genomics in mammalian cells: haploid mouse embryonic stem (ES) cells.[5] The types of haploid organisms—those carrying single copies of each chromosome—range from yeast to social insects and fish. There are also near-haploid human tumor cells. Insects, fish, and near-haploid tumor cells have proven to be very useful for carrying out genome-wide mutagenesis studies and for analyzing recessive phenotypes.[6,7] Penninger reported the generation of the first haploid mouse ES cell lines, which carry 20 chromosomes, express bona fide stem cell markers, and maintain genome integrity. Functionally, the haploid ES cells can develop, *in vitro* and *in vivo*, into cell types of all germ layers; the haploid ES cells can be readily mutagenized, thus making it possible to perform whole-genome forward genetics.

Since all ES cells offer access to unlimited quantities of nearly every cell type, haploid ES cells may provide a tool to genetically assess fundamental developmental and biological processes in defined cell types found in adipocyte differentiation or to clarify functions of mature white, brite, and brown adipocytes.

Precursors and secular trends in metabolic disease

David I.W. Phillips (University of Southampton) discussed many of the factors driving global trends of increasing incidence of metabolic disease, and the mechanisms underlying the increased susceptibility of populations vulnerable to these health risks. Type 2 diabetes and related metabolic diseases have become epidemic in several resource-poor countries that have undergone rapid westernization.[8] Although this epidemic was precipitated by the abandonment of traditional lifestyles, consumption of energy-rich foods, reduced physical activity, and increased obesity, these are not the only factors involved in high disease rates in these countries. People with a history of poverty and undernutrition may be uniquely susceptible to these lifestyle changes. Recent research suggests that the process of developmental plasticity, also referred to as early life

programming, may offer a cogent explanation of this susceptibility.

Developmental plasticity describes a process by which organisms (or populations) adapt to adverse environments by developing nongenetically determined phenotypic alterations in body composition and physiology to counteract the adversity. However, such changes can only be considered adaptive (and thus healthy) if the population continues to live in the conditions responsible for the adverse environment (e.g., poverty). The adaptations are (or can become) maladaptive (unhealthy), in contrast, when the environmental adversity is (externally) removed, such as when a poor country rapidly westernizes.[9]

Developmental plasticity is present in most animal species and has been well documented in animal experiments. In human studies, low birth weight and poor postnatal growth have been used as proxies for developmental adversity. Low birth weight, prevalent in many resource-poor countries, is associated with disturbed body composition—central obesity, low muscle mass, and poor bone mineralization—in later life. It is also linked to abnormalities of carbohydrate metabolism (insulin resistance and defective insulin secretion), vascular function (high blood pressure), and functions of major organs such as the liver and kidneys; these abnormalities raise the risk of type 2 diabetes, hypertension, and cardiovascular disease. Current studies of mechanisms underlying such phenotypic and physiological alterations imply that they are epigenetic. Some pathways have been defined. For example, early life stress is known to alter the epigenetic regulation of glucocorticoid receptors in target tissues; this results in lifelong changes in the hormonal systems that mediate the biobehavioral response to stress, a known regulator of carbohydrate metabolism and vascular physiology.[10] It follows that improving the nutrition of pregnant women and infants—to reduce the likelihood of unhealthy alterations in body composition later—may be an effective way to prevent these emerging health problems in some developing countries.

Sarcopenia and muscle power

Sarcopenia: normal versus pathologic

Steven B. Heymsfield (Louisiana State University) shifted the focus to examine aging-related diseases. A long-recognized age-related phenomenon, sarcopenia— the gradual aging-related loss of skeletal muscle mass with associated changes in muscle quality and function[11]—is a clinically important component of frailty and some metabolic disturbances. Sarcopenia is the focus of active research programs aimed at understanding underlying mechanisms in order to develop preventive and therapeutic measures. Gradual loss of skeletal muscle mass is a normal part of the aging process, related to increasing inactivity. It can also be caused by underlying catabolic illness, usually referred to as *cachexia*, which differs metabolically from classical sarcopenia.[11] Thus, sarcopenia is classified as primary or secondary (inactivity versus cachexia); primary sarcopenia appears in stages, beginning with pre-sarcopenia, advancing to clinically manifest sarcopenia, and finally to severe sarcopenia.[11]

Recently sarcopenia has been recognized in obese individuals, a comorbidity referred to as *sarcopenic obesity*. As might be expected from two conditions that individually pose increased risk, sarcopenic obesity is associated with greater morbidity risk than either condition alone.

Skeletal muscle mass reaches peak values during the late teen years and early twenties, and then begins to slowly decline in healthy adults at a rate of about 0.5–1% per year. Age-related rates of skeletal muscle mass loss vary; individuals with rapid muscle loss are at risk for sarcopenia, as are those with rapid bone loss and osteoporosis. Determinants of skeletal muscle mass include genetic susceptibility, height, activity level, race, adiposity, and abnormal levels of key hormones.

Heritability estimates for human skeletal muscle (lean) mass is ~ 0.52, with similar corresponding estimates for leg extensor and grip strength.[12] Myostatin is a gene that regulates skeletal muscle mass; inactivation leads to significantly larger skeletal muscles in mammals. Taller adults have more skeletal muscle than their shorter counterparts. Skeletal muscle scales with the square of height, similar to body weight.[13] Analogous to body mass index (BMI; weight/height^2), *skeletal muscle mass index* (SM/height^2) adjusts for individual differences in height, allowing for the definition of diagnostic criteria for sarcopenia. People who exercise regularly have greater skeletal muscle mass. A large percentage of Americans have suboptimal levels of leisure time physical activity, particularly the elderly. On average, African Americans have more skeletal muscle

mass than Caucasians matched for weight, height, and age,[14] through unknown mechanisms. Anabolic hormones, such as androgens, growth hormone, and IGF-1, stimulate growth and maintenance of skeletal muscle mass, effects that decline with aging, particularly from the sixth decade onward. Finally, greater skeletal muscle mass develops with larger mechanical loads. Thus, obese people tend to have more skeletal muscle than age-matched counterparts; the proportion of fat-free mass as skeletal muscle increases with greater adiposity.

While the classical approach to sarcopenia focuses on skeletal muscle mass, growing interest surrounds *dynapenia*, or loss in muscle strength.[11] Functional limitations, frailty, and other consequences of age-related changes in muscle strength are of central importance to the morbidity and mortality associated with sarcopenia. Beginning around the seventh decade and progressing beyond, rates of loss of skeletal muscle function exceed those of mass. Functional effects are very slow to return following periods of illness and lack of activity.

Obviously, sarcopenia results from multiple interacting factors, the mechanisms of which remain to be determined. Skeletal muscle does not develop in isolation; it grows during development and declines with senescence under the influence of neural and hormonal regulation. These factors influence skeletal muscle mass and function and contribute to the pathogenesis of sarcopenia. For example, an elderly person's fall with injury may be attributable to insufficient skeletal muscle mass with poor functional quality (poor leg strength and coordination), as well as to a series of neurocognitive and neuromuscular pathways functioning below optimal levels. Thus, it is only within a broader physiological context that a full understanding of sarcopenia and its clinical manifestations will evolve.

Skeletal muscle power: a determinant of physical functioning in older adults

Roger A. Fielding (Tufts University) discussed the importance of muscle power output as one of the variables that contribute to physical impairment associated with aging. Skeletal muscle, the largest organ mass in the body, making up 40–50% of total body mass, is required for locomotion and is a determinant of oxygen consumption, whole body energy metabolism, and substrate turnover and storage. Advancing age is associated with the loss of skeletal muscle (sarcopenia), and can lead to declines in physical functioning. The causes of sarcopenia, like many complex geriatric syndromes, are multifactorial.[15] The loss of muscle mass and strength, and the concurrent increase in joint dysfunction and arthritis that occurs with aging, results in a decrease in physical function and an increased risk of disability.[16] Disability is associated with limitations in performing regular activities of daily living, including rising from a chair, climbing flights of stairs, bathing, and preparing food. Studies have shown that the ability to perform these tasks decreases with age in both men and women.[17]

Despite the high prevalence and major health implications, sarcopenia still has no broadly accepted clinical definition or diagnostic criteria. The most current definition of sarcopenia includes gait speed < 1.0 m/s combined with a low ratio of appendicular lean mass (aLM) to height squared (≤ 7.23 kg/m^2 in males, ≤ 5.67 kg/m^2 in women),[18] two standard deviations below the mean aLM of young healthy adults.[19]

Skeletal muscle power output (the product of force and velocity) declines earlier and more precipitously with advancing age compared to muscle strength. In cross-sectional and longitudinal studies of older adults, Fielding and colleagues observed that declines in muscle mass could not fully explain the observed deficits in skeletal muscle power output, suggesting that factors related to deficits in neuromuscular activation may play a role.[20] Fielding reported significant reductions in neuromuscular activation patterns and deficits in rapid activation of voluntary lower extremity muscle groups in older adults with measured limitations in physical functioning. Peak muscle power has also emerged as an important predictor of functional limitations in older adults, and as an important outcome measure in clinical trials of resistance training of older adults.[21] Fielding explained that his group's current working hypothesis is focused on examining lower extremity muscle power as a more critical variable in understanding the relationship between impairments, functional limitations, and resultant disability with aging.

Cognition and diabetes

Cognitive reserve and aging

Yaakov Stern (Columbia University College of Physicians and Surgeons) discussed the concept of

reserve—a variable that refers to the discrepancy (or discordance) between brain pathology and clinically apparent symptoms. Stern explained that reserve refers to differences between two components: brain reserve is the number of neurons or synapses unaffected by pathology; cognitive reserve is the resilience or plasticity of cognitive networks compensating for loss of functional neurons. Stern's group studies cognitive reserve using neuroimaging.

Whatever the mechanism, reserve is malleable and can strengthen or weaken at every stage of life. In AD and other diseases of aging associated with cognitive decline, the initiating and promoting pathology can precede the onset of clinical symptoms by decades, depending on reserve. Exercise and environmental factors stimulate brain plasticity and can remodel neuronal circuitry, which in turn increases vascularization and neurogenesis in the dentate, and thus brain volume, cortical thickness, neuronal survival, and resistance to progressive brain pathology.

The concept of reserve has been supported by epidemiological studies. After following a cohort of clinically asymptomatic elderly individuals to determine the time of onset of AD symptoms,[22] Stern and colleagues report that individuals grouped by education level have different outcomes; those with less formal education were twice as likely to develop AD as those with more. Similarly, individuals in occupations with lower cognitive demands were at twice the risk of AD as those with high occupational attainment, and greater demands. Among the same cohort, when controlled for education and occupational attainment, individuals who engaged in high levels of physical activity had reduced risk of AD compared with those with low activity. In studies of elderly individuals without diagnosed AD, higher levels of literacy correlated with lower rates of memory decline.[23]

Reserve seems to offer significant but incomplete protection against AD and other brain pathologies: individuals with high reserve and little pathology may be free of dementia or other clinical signs, while individuals with high reserve and moderate pathology may have less severe dementia than individuals with low reserve. This suggests that individuals with low reserve and mild severity likely have less pathology. In imaging studies controlled for clinical disease severity, Stern has reported an inverse relationship between level of education and parietotemporal perfusion deficit, a marker of AD.[24]

In more recent studies, Stern's group used imaging techniques (e.g., fMRI) to examine brain regions that may be involved in reserve. A letter-recognition test unveiled two active brain networks: one primary, used by both young and elderly subjects, and one exclusively used by the elderly. While this suggested a compensatory neural network, individuals who used the second network performed comparatively poorly on the letter recognition task, and the more atrophy present within an individual's primary network the more likely she would be to use the second network. However, Stern showed that, given a level of atrophy, individuals with higher cognitive performance (IQ) performed better on the letter recognition test, even among those who used the second network.[25]

To summarize, Stern reiterated how the concept of reserve is supported by epidemiological and imaging evidence, and how it appears malleable, even at later stages of life. Reserve may be useful in a range of conditions affecting brain function. Influencing reserve has the potential to delay or reverse aging and brain pathology.

Diabetes and the brain

Lenore J. Launer (National Institute on Aging) provided an overview of recent reports demonstrating greater cognitive impairment in type 2 diabetes (T2D), associated with underlying structural changes with functional consequences. Obesity has many comorbidities, the most prevalent being diabetes. The rise in the prevalence and incidence of T2D and obesity[26] is reaching epidemic proportions, with both conditions occurring earlier in life. Another public health concern nearing epidemic levels is the logarithmic increase in the incidence of dementia after 65 years of age. There have been several reports of associations between obesity and dementia, and between dementia and T2D in the past decade.[27] Although data linking BMI and dementia are inconsistent (see Gustafson, below), the relationship between T2D and dementia is fairly robust, partly explaining the association between obesity and dementia.

Diabetes is associated with both vascular disease and neurodegenerative processes associated with cognitive impairment. These oxidative, inflammatory, metabolic, protein-misfolding, and neuroendocrine processes connect diabetes with mixed brain pathology that includes large, small, and microinfarcts and AD lesions (neuritic plaques, neurofibrillary tangles, and cerebral amyloid angiopathy).

Several epidemiologic studies examining several ethnic groups have reported approximately twice the risk for dementia, including AD, in individuals with T2D compared to those without diabetes. Evidence for the link between diabetes and dementia is further drawn from studies of associations between T2D and intermediary measures of brain structure and function.[27–29] They suggest that individuals with T2D perform more poorly on tests of memory, processing speed, and executive function, and that performance worsens with duration of diabetes and poor glycemic control. Magnetic resonance imaging (MRI) demonstrates that patients with T2D have increased numbers of infarct-like lesions, hippocampal atrophy (reflecting neurodegeneration), or both. There is additional indirect evidence that microvascular pathology is associated with poor performance on speed and executive function tests, and that diabetic patients have cerebral microvascular angiopathy, MRI-detected white matter damage, and cerebral microbleeds. In addition, patients with T2D have reduced total brain and gray matter volume, suggesting that, in addition to neurodegeneration, loss of gray matter microvasculature contributes to the brain atrophy seen in dementia.

Finally, genetic susceptibility factors seem to modulate the risk of cerebral pathology in individuals with T2D. Autopsy studies in the Honolulu-Asia Aging Study suggest that those with diabetes and the apolipoprotein E ε4 allele, a genetic susceptibility factor for AD, have a much higher risk for cerebral amyloid angiopathy, infarcts, and neurotic plaques than those without diabetes and the ε4 allele. A similar association was found for clinical dementia.[27] With rapidly evolving genetic technology, it is likely that such links will be better understood in the next few years.

Although there is strong evidence that dysglycemia in itself can contribute directly and indirectly to vascular damage and neurodegeneration, T2D patients and those with the metabolic syndrome, including insulin resistance, hypertension, coronary disease, and dyslipidemia, often exhibit brain pathology.[28] Studies of the association of T2D with brain structure/function need to include models adjusted for these comorbidities and to test for compounding effects of these comorbidities to determine pathology.

Measuring and modifying risks

The effects of obesity on dementia

Deborah R. Gustafson (SUNY Downstate Medical Center, and the University of Gothenburg) discussed the relationship between body mass and risks for AD. Body weight and BMI are common, simple measures of overnutrition, commonly termed *overweight* and *obesity*. BMI and central obesity, measured as waist circumference (WC) or waist-to-hip ratio (WHR), are linked to manifest dementia and late-onset AD (LOAD) in epidemiological studies.[30–33] Overweight and obesity in middle and later life may increase risk for dementia, whereas decline in body weight or BMI and underweight in years preceding and at the time of LOAD diagnosis may also contribute to dementia and its clinical progression.

While midlife (adult) and late-life (approximately 60 years and older) factors are associated with AD, the pathogenesis is more complex. On average, BMI increases throughout life until age 60–70 years; higher midlife BMI or central obesity decades before dementia onset are linked to higher risk of AD later in life.[34–38] Risk estimates for AD and all dementias associated with overweight and obesity are in the range of 1.5- to 3-fold higher than among normal-weight individuals. This level of risk is similar to that observed for hypertension and other cardiovascular risk factors. Levels of midlife BMI and central obesity associated with AD are in overweight and obese ranges, based on traditional cutoffs used in majority Caucasian populations for assessing cardiovascular risk and predicting overall mortality (e.g., BMI 25 kg/m^2 or WHR 0.85 in women and 0.90 in men). Subsequently, BMI declines. Thus, while high levels of BMI during midlife may increase risk for the chronic neurodegenerative diseases of aging, the direction of the BMI–AD relationship appears to plateau and/or change in late life.[38,39] On average, in later life individuals with AD have a lower body weight or BMI than those without AD. This paradoxical combination of higher LOAD risk associated with midlife overweight and obesity and decline in BMI and underweight in the years immediately preceding and at the time of LOAD diagnosis requires further investigation. Ongoing studies are evaluating adipose tissue hormones (adipokines), adipose tissue vascularity and angiogenesis, and regional differences between adipose depots in relation to clinical dementia and brain structure and function.

Nutrition and cognition: limitations, complexities, and interpretations

Nikolaos Scarmeas (University of Athens and Columbia University) discussed the limitations of, and complexities in, our understanding of the links between nutrition and cognition. Studies have found that cognitive performance and risk for AD and other brain diseases associated with dementia vary with intake of micro- and macronutrients and with dietary pattern. And while this literature is extensive, studies are often not confirmed and associations are reported that are not always consistent.[40,41] As a result, few clear recommendations can be made regarding nutritional habits with the potential to improve cognitive performance or protect from dementia or cognitive impairment. This uncertainty might engender the idea that no definitive association exists between nutrition and cognition. Scarmeas urged, however, that methodological limitations are the root of the problem. For example, nutritional supplementation may only be effective in people with actual deficiencies, in contrast to those who participate in scientific studies and have a normal nutritional status, leaving little possibility of physiological improvement.[42]

Scarmeas also described difficulties in measuring cognitive outcomes, including the subjectivity of dementia diagnosis, a lack of reliable biomarker diagnosis, and the variability of neuropsychological evaluations. There are also significant inherent limitations in the assessment of nutritional exposure. For example, food frequency questionnaires probe for answers that require complex consideration or focus on a limited number of foods covering a proportion of overall diet. Nutrient biomarkers can be inaccurate, costly, and usually do not indicate central nervous system levels of nutrients; they are also limited to a relatively low number of nutritional elements, while a normal diet includes thousands of chemicals, many of which have cognitive effects. Furthermore, self-reporting is notoriously inaccurate. Analyses that follow from a reductionist consideration of a limited number of nutritional elements impose the significant constraint of ignoring both confounders and agents, while simultaneously failing to summarize dietary habits in a single measure. Scarmeas urged a more holistic approach to dietary patterns, which would partially remedy some of these problems and provide significant public health information, even though not elucidating mechanisms explaining interactions between diet and cognition.[43]

Finally, cognition is an extremely complex activity influenced by a multiplicity of cerebral biological mechanisms (including vascular, inflammatory, metabolic, oxidative, and β-amyloid-, τ-, and α-synuclein–related). Consequently, there are likely multiple nutritional effects with interacting, and even opposing, directionality.[44]

A series of attempts to partially address some of the above issues includes consideration of baseline levels of nutrients, measurements of nutrient biomarkers and examination of their relationship with neurodegeneration-related (and other mechanism-related) biomarkers,[44] estimates of dietary patterns,[42] and the use of brain imaging biomarkers of various types. Although such attempts have emerged only relatively recently, they may soon increase in number and frequency, enhancing the current limited understanding of the relationship between nutrition and cognition.

Bariatric surgery, cognitive function, and Alzheimer's disease

Further discussing the links between obesity and cognitive dysfunction, John Gunstad (Kent State University) presented an analysis of the beneficial metabolic effects of bariatric surgery on cognitive function and AD. Gunstad emphasized that while obesity has been recognized as an independent risk factor for adverse neurocognitive outcomes, including AD, vascular dementia, stroke, and accelerated cognitive decline, it is also associated with cognitive dysfunction long before onset of these conditions. Deficits in attention, executive function, memory, and psychomotor speed are commonly found on neuropsychological tests done before the occurrence of overt symptoms of neurodegenerative disease. Such deficits accord with neuroimaging findings in the obese population, including global and specific atrophy, reduced frontal lobe metabolism, and white matter abnormalities (reviewed in Ref. 45). In the absence of evidence that these types of neuropathology might be reversible, it was tempting to study whether obesity-related cognitive dysfunction could be reversed through significant weight loss.

Obesity is associated with many conditions exhibiting reversible cognitive deficits, including hypertension, type 2 diabetes, sleep apnea, and depression, which support the hypothesis that weight loss may improve cognitive function and ultimately reduce risk of adverse neurocognitive outcomes in obese patients.[45] An ongoing project is examining this possibility by prospectively assessing cognitive function in patients that undergo various forms of bariatric surgery. Bariatric operations are safe and effective for metabolic obesity, and gastric bypass patients can lose more than 60% of their excess weight at nadir.[46] If weight loss were to produce improvements in cognitive function, it would most likely appear in severely obese or dysmetabolic individuals who maintain a medically significant amount of weight loss.

Gunstad described a prospective assessment[47] of 125 patients undergoing either Roux-en-Y gastric bypass or gastric banding from the Longitudinal Assessment of Bariatric Surgery (LABS) parent project, as well as 125 demographically similar obese control patients. As predicted, cognitive impairment was prevalent in bariatric surgery candidates, with nearly 25% exhibiting clinically significant levels of dysfunction (>1.5 SD below normative performance) prior to surgery; up to 40% exhibited milder deficits (1 SD below normative performance). Nearly all patients underwent Roux-en-Y gastric bypass procedures, and those without surgical complications did not exhibit cognitive decline at 12 weeks postoperative. While past studies show that a small number of patients experience neurological complications due to nutritional deficiencies over the long term after surgery,[48] often owing to failure to adhere to supplementation, the absence of widespread cognitive dysfunction on testing in this ongoing project suggests these complications are fairly rare.

More interesting were findings that surgical patients exhibited significant improvements compared to baseline on multiple cognitive tests as early as 12 weeks postoperative. Most notably, performance improved on multiple memory indices (e.g., learning, recall, recognition) in surgery patients, whereas memory declined in obese controls during the same relatively short period of time.[47] *Post hoc* analyses showed that magnitude of weight loss and improvement in comorbid medical conditions contributed very little to this effect.

Continued assessment of these patients reveals that the cognitive benefits of bariatric surgery persist over time. A series of submitted papers describe that bariatric surgery patients show further cognitive improvement one year postoperative and maintenance of the gains at two years. Further studies are needed to determine whether cognitive benefits persist for longer periods (e.g., 60 months) when weight regain is common among patients undergoing any form of weight loss surgery, and the degree to which these procedures may reduce risk for AD or other neurological disorders later in life.

Conclusions

The material presented here summarizes the current understanding of the risks posed by overnutrition for cognitive decline, Alzheimer's disease, dementia, sarcopenia, and other diseases associated with aging. Crucially, the risks of overnutrition are potentially modifiable; this is especially significant if addressing overnutrition can minimize or mitigate the development of dementia, for which no effective treatments exist and monetary cost in the United States alone reaches over $150 billion per year.[49] Effectively reducing these risks will require further exploration of appropriate and optimal temporal targets for intervention—whether the obese brain is entrained very early in life or whether the potential for exercise and neural plasticity to reduce risks of cognitive decline persists throughout life. Improvements in epidemiological methods will be necessary to fully examine the relationship between obesity and cognitive decline, including technological advances in telemedicine that can standardize and reduce the irregular intervals at which epidemiological data are collected. Large cohorts of patients tracked from birth are now reaching ages relevant to the study of these conditions, and data from these studies may be essential to understanding the contribution of childhood factors to diseases of aging. Developing work linking more sophisticated measures of body composition, hormones, and metabolic markers to specific subpopulations with genetic polymorphisms may yield more definitive results than past work correlating outdated measures like BMI with clinical outcomes.

Ultimately, however, the greater challenge in addressing the relationship between nutrition, lifestyle, and aging lies in translating and implementing what is already known into interventions and preventive measures in the population. The relationships between overnutrition and diseases like diabetes and cardiovascular disease are known; however, this knowledge has not reduced the trends toward increasing overweight and obesity in the United States and many other countries. Still, a reasonable approach to address the risks for cognitive decline and sarcopenia, as well as other diseases of obesity, is education at the public, academic, and political levels in order to ignite dynamic changes in the priorities of public policy, the paradigms of research funding, and the messaging and communications of motives to both physicians and patients.

Acknowledgments

The conference "The Paradox of Overnutrition in Aging and Cognition" was supported by the Sackler Institute for Nutrition Science at the New York Academy of Sciences.

Conflicts of interest

The authors declare no conflicts of interest.

References

1. Ashrafi, K., F.Y. Chang, J.L. Watts, *et al.* 2003. Genome-wide RNAi analysis of Caenorhabditis elegans fat regulatory genes. *Nature* **421:** 268–272.
2. Dietzl, G., D. Chen, F. Schnorrer, *et al.* 2007. A genome-wide transgenic RNAi library for conditional gene inactivation in *Drosophila. Nature* **448:** 151–156.
3. Pospisilik, J. A., D. Schramek, H. Schnider, *et al.* 2010. *Drosophila* genome-wide obesity screen reveals Hedgehog as a determinant of brown versus white adipose cell fate. *Cell* **140:** 148–160.
4. Teperino, R., S. Amann, M. Bayer, *et al.* 2012. Hedgehog partial agonism drives Warburg-like metabolism in muscle and brown fat. *Cell* **151:** 414–426.
5. Elling, U., J. Taubenschmid, G. Wirnsberger, *et al.* 2011. Forward and Reverse Genetics through Derivation of Haploid Mouse Embryonic Stem Cells. *Cell Stem. Cells* **9:** 563–574.
6. Carette, J.E., C.P. Guimaraes, M. Varadarajan, *et al.* 2009. Haploid genetic screens in human cells identify host factors used by pathogens. *Science* **326:** 1231–1235.
7. Streisinger, G., C. Walker, N. Dower, *et al.* 1981. Production of clones of homozygous diploid zebra fish (Brachydanio rerio). *Nature* **291:** 293–296.
8. Fall, C.H. 2004. Fetal Programming and the Risk of Noncommunicable Disease. *Indian J. Pediatr.* **S1:** 13–20.
9. Bateson, P., D. Barker, T. Clutton-Brock, *et al.* 2004. Developmental plasticity and human health. *Nature* **430:** 419–421.
10. Phillips, D.I. 2007. Programming of the stress response: a fundamental mechanism underlying the long-term effects of the fetal environment? *J. Intern. Med.* **261:** 453–460.
11. Cruz-Jentoft, A.J., J.P. Baeyens, J.M. Bauer, *et al.* 2010. Sarcopenia: European consensus on definition and diagnosis:

Report of the European Working Group on Sarcopenia in Older People. *Age Ageing.* **39:** 412–423.

12. Arden, N.K. & T.D. Spector. 1997. Genetic influences on muscle strength, lean body mass, and bone mineral density: a twin study. *J. Bone Miner. Res.* **12:** 2076–2081.

13. Heymsfield, S.B., M. Heo, D. Thomas & A. Pietrobelli. 2011. Scaling of body composition to height: relevance to height-normalized indexes. *Am. J. Clin. Nutr.* **93:** 736–740.

14. Heymsfield, S.B., R. Scherzer, A. Pietrobelli, *et al.* 2009. Body mass index as a phenotypic expression of adiposity: quantitative contribution of muscularity in a population-based sample. *Int. J. Obes.* **33:** 1363–1373.

15. Morley, J.E., R.N. Baumgartner, R. Roubenoff, *et al.* 2001. Sarcopenia. *J. Lab. Clin. Med.* **137:** 231–243.

16. Jette, A.M., S.F. Assmann, D. Rooks, *et al.* 1998. Interrelationships among disablement concepts. *J. Gerontol.* **53A:** M395–M404.

17. Jette, A.M. & L.G. Branch. 1981. The Framingham Disability Study: II. Physical disability among the aging. *Am. J. Public Health.* **71:** 1211–1216.

18. Fielding, R.A., B. Vellas, W.J. Evans, *et al.* 2011. Sarcopenia: an undiagnosed condition in older adults. Current consensus definition: prevalence, etiology, and consequences. International working group on sarcopenia. *JAMDA* **12:** 249–256.

19. Baumgartner, R.N., K.M. Koehler, D. Gallagher, *et al.* 1998. Epidemiology of Sarcopenia among the elderly in New Mexico. *Am. J. Epidemiol.* **147:** 755–763.

20. Clark, D.J. & R.A. Fielding. 2012. Neuromuscular contributions to age-related weakness. *J. Gerontol. A. Biol. Sci. Med. Sci.* **67:** 41–47.

21. Reid, K.F. & R.A. Fielding. 2012. Skeletal muscle power: a critical determinant of physical functioning in older adults. *Exercise Sport Sci. R.* **40:** 4–12.

22. Stern, Y., B. Gurland, T.K. Tatemichi, *et al.* 1994. Influence of Education and Occupation on the Incidence of Alzheimer's Disease. *JAMA.* **271:** 1004–1010.

23. Manly, J.J., P. Touradji, M.X. Tang & Y. Stern. 2003. Literacy and memory decline among ethnically diverse elders. *J. Clin. Exp. Neuropsychol.* **25:** 680–690.

24. Stern, Y., G.E. Alexander, I. Prohovnik & R. Mayeux. 1992. Inverse relationship between education and parietotemporal perfusion deficit in Alzheimer's disease. *Ann. Neurol.* **32:** 371–3765.

25. Steffener, J. & Y. Stern. 2012. Exploring the neural basis of cognitive reserve in aging. *Biochim. Biophys. Acta.* **1822:** 467–473.

26. Flegal, K.M., M.D. Carroll, B.K. Kit & C.L. Ogden. 2012. Prevalence of obesity and trends in the distribution of body mass index among US adults, 1999–2010. *JAMA* **307:** 491–497.

27. Launer, L.J. 2009. Diabetes: vascular or neurodegenerative: an epidemiologic perspective. *Stroke* **40:** S53–55.

28. Gorelick, P.B., A. Scuteri, S.E. Black, *et al.* 2011. American Heart Association Stroke Council, Council on Epidemiology and Prevention, Council on Cardiovascular Nursing, Council on Cardiovascular Radiology and Intervention, and Council on Cardiovascular Surgery and Anesthesia. Vascular contributions to cognitive impairment and dementia: a statement for healthcare professionals from the ameri-can heart association/american stroke association. *Stroke* **42:** 2672–2713.

29. Saczynski, J.S., S. Siggurdsson, P.V. Jonsson, *et al.* 2009. Glycemic status and brain injury in older individuals: the age gene/environment susceptibility-Reykjavik study. *Diabetes Care* **32:** 1608–1613.

30. Gustafson, D.R. 2006. Adiposity indices and dementia. *Lancet Neurol.* **5:** 713–720.

31. Gustafson, D.R., E. Rothenberg, K. Blennow, *et al.* 2003. An 18-year follow up of overweight and risk for Alzheimer's disease. *Arch. Intern. Med.* **163:** 1524–1528.

32. Gustafson, D. 2008. A life course of adiposity and dementia. *Eur. J. Pharmacol.* **585:** 163–175.

33. Anstey, K.J., N. Cherbuin, M. Budge & J. Young. 2011. Body mass index in midlife and late-life as a risk factor for dementia: a meta-analysis of prospective studies. *Obes. Rev.* **12:** e426–437.

34. Whitmer, R.A., E.P. Gunderson, C.P. Quesenberry, Jr., *et al.* 2007 Body mass index in midlife and risk of Alzheimer disease and vascular dementia. *Curr. Alzheimer Res.* **4:** 103–109.

35. Whitmer, R.A., D.R. Gustafson, E. Barrett-Connor, *et al.* 2008. Central obesity and increased risk of dementia more than three decades later. *Neurology* **71:** 1057–1064.

36. Fitzpatrick, A.L., L.H. Kuller, O.L. Lopez, *et al.* 2009. Midlife and late-life obesity and the risk of dementia: cardiovascular health study. *Arch. Neurol.* **66:** 336–342.

37. Kivipelto, M., T. Ngandu, L. Fratiglioni, *et al.* 2005. Obesity and vascular risk factors at midlife and the risk of dementia and Alzheimer disease. *Arch. Neurol.* **62:** 1556–1560.

38. Gustafson, D.R., K. Bäckman, M. Waern, *et al.* 2009. Adiposity indicators and dementia over 32 years in Sweden. *Neurology* **73:** 1559–1566.

39. Gustafson, D., K. Bäckman, E. Joas, *et al.* 2012. A 37-year longitudinal follow-up of body mass index and dementia in women. *J. Alzheimers Dis.* **28:** 162–171.

40. Daviglus, M.L. *et al.* 2011. Risk factors and preventive interventions for Alzheimer disease: state of the science. *Arch. Neurol.* **68:** 1185–1190.

41. Plassman, B.L. *et al.* 2010. Systematic review: factors associated with risk for and possible prevention of cognitive decline in later life. *Ann. Int. Med.* **153:** 182–193.

42. Morris, M.C. & C.C. Tangney. 2011. A potential design flaw of randomized trials of vitamin supplements. *JAMA* **305:** 1348–1349.

43. Gu, Y. & N. Scarmeas. 2011. Dietary patterns in Alzheimer's disease and cognitive aging. *Curr. Alzheimer Res.* **8:** 510–519.

44. Galasko, D.R. *et al.* 2012. Antioxidants for Alzheimer disease: a randomized clinical trial with cerebrospinal fluid biomarker measures. *Arch. Neurol.* **69:** 836–441.

45. Stanek, K. & J. Gunstad. 2012. Can bariatric surgery reduce risk of Alzheimer's disease? *Prog. Neuropsychopharmacol. Biol. Psychiatry*; online July 2012.

46. Buchwald, H., Y. Avidor, E. Braunwald, *et al.* 2004. Bariatric surgery: a systematic review and meta-analysis. *JAMA* **292:** 1724–1737.

47. Gunstad, J., G. Strain, M.J. Devlin, *et al.* 2011. Improved memory function 12 weeks after bariatric surgery. *Surg. Obes.Relat. Dis.* **7:** 465–472.

48. Becker, D., L. Balcer & S. Galetta. 2012. The neurological complications of nutritional deficiency following bariatric surgery. *J. Obes* 608534.

49. Hurd, M.D., P. Martorell, A. Delavande, *et al.* 2013. Monetary costs of dementia in the United States. *N. Eng. J. Med.* **368:** 1326–1334.

Synopsis and commentary

John G. Kral

Synopsis

The problem

The proportion of the U.S. population over age 65 grew from 9.8% in 1970 to 12.4% in 2000 (http://www.cdc.gov/nchs/data/databriefs/db91.htm) and is expected to reach 25% by 2030. Morbidity rises with age, and rising longevity increases demands for medical care and assisted living. Although overall chronic disability prevalence in the aging population has been declining, the proportion of the population in assisted living continues to increase, as does the prevalence of cognitive impairment. One in three seniors dies from dementia.

The root cause

As with global warming, the rapid rise of civilization has driven a shift in biological energy balance from homeostatic (or eumetabolic) to hypercaloric, for which *Homo sapiens* is not yet adapted. Progress has enabled greater availability and improved distribution, processing, and preservation of food in step with labor-saving technological advances, and urban living in the built environment, infection control, hygiene, public health, and medical care have all contributed to, and been credited for, the centuries-long trend of increasing longevity. The appearance and acceleration of civilization-related diseases during the last 70 years seem to be threatening this progress: increasing longevity in industrialized nations has plateaued[1] and deaths from noncommunicable disease have increased world-wide.[2]

The paradox

Relationships between aging and cognitive decline, frailty, weight-loss (or undernutrition), and established risk factors such as hypertension, diabetes, cardiopulmonary failure, and dyslipidemia have been well-known for decades. Owing to its prevalence and its burgeoning share of medical costs, the chronic inflammatory insulin-resistant overnutrition syndrome, traditionally termed *obesity*, has only recently been acknowledged as a serious disease related to the comorbidities of aging, indeed suggesting that obesity is a form of accelerated aging.[3] This meeting was convened to address the apparent paradox or problem of reconciling concurrent adverse effects of overnutrition (obesity) in early life and midlife with progressive terminal undernutrition.

Pathways to solutions

The symposium uniquely brought together experts in molecular biology, endocrine and metabolic programming, body composition, exercise and integrative physiology, epidemiology of diabesity and neurology, and functional neuroimaging, enabling conversations among scientists who do not normally interact. The presentations provided insights into (1) critical mechanisms of adipogenesis, early-life stress, and sarcopenia together contributing to cognitive decline, (2) methods for assessing sarcopenia, muscle physiology, and neurocognitive function, and (3) preventive strategies for preserving or even restoring cognitive function. Although several important areas were not represented, the meeting provided constructive focus for a research agenda.

Commentary

Mismatch between symptoms and pathology

The diverse interdisciplinary backgrounds of the speakers highlight the need to clearly define terms central to the topic of the meeting. Obesity (from Latin: overeat) is very easy to diagnose; it is commonly described in terms of phenotypic size using weight and height to construct the body mass index (BMI), which generally follows a continuous normal distribution. Even though BMI is continuous, discrete cut-off levels of BMI are used to define underweight, overweight, and obesity (without correcting for age, sex, or race), equating them with morbidities such as undernutrition and overnutrition, although the dysmetabolic manifestations associated with tissue and organ impairment do not agree with the cut points.[4] In this regard, there is a mismatch between the signs and symptoms and the pathology. The time has come to replace *obesity* as a catchall for a syndromic dysmetabolic condition related to positive energy balance by *overnutrition*, and to replace BMI with more appropriate disease-related biomarkers such as waist circumference, plasma C-reactive protein (CRP), glycated hemoglobin (HbA1c), HDL-cholesterol and leptin, and analysis of microalbuminuria.

Similarly, *Alzheimer's disease* is commonly used to denote a decline in cognitive function with defined phenotypic features of dementia, although AD is a specific abnormality of subcellular and cellular structure of brain tissue associated with functional changes ultimately manifesting as diagnosable cognitive impairment. Originally, Alzheimer described characteristic histopathological brain changes in patients with advanced manifest dementia during an epoch with significantly shorter life span than the current. As life spans increased, pressure increased to detect early signs and symptoms with the hope of preventing the disease. It is only recently that robust early biomarkers have become available.[5] This will enable the design of studies of preventive measures.

Challenges

The ultimate challenge for preventing and treating aging-related dementia is germane to all comorbidities associated with the dysmetabolic chronic overnutrition syndrome of obesity: sustaining or achieving a eumetabolic isoenergetic state. Obesity is indeed easy to diagnose. There are easily identifiable risk factors with great predictive power such as family history, race, sedentism, diet, occupation, and many more.[6] Nevertheless, the long lead time from the start of the disease processes, especially evident with cognitive decline and dementia, to appearance of disease above the clinical horizon poses especially difficult diagnostic and treatment challenges in childhood.

Recognition of the huge importance of the intrauterine environment as a primary locus for establishing precursors of adult disease (developmental programming; fetal origins) should prompt policies and procedures to ameliorate the often neglected environments of conception and gestation. The barriers are tremendous. Theoretically, pregnancy planning supported by education should be prioritized and well supported in civilized, enlightened communities. It is often argued that money is not a sufficient remedy for poverty-related problems. However long-term planning is not an option for those struggling with the stresses of day-to-day survival and adverse early-life experiences determine later morbidity and mortality.[7]

Problems and promises of plasticity

Just as developmental plasticity on a population and organismic level can be blamed for maladaptation and the ills highlighted in this meeting, plasticity has great positive value throughout the life span, thus providing reasons for optimism. Accumulating evidence affirms treatment and prevention benefits starting in the womb and continuing into advanced senescence. Tissue remodeling of heart and vessels and neurogenesis late in life support observations that judicious training of muscles (exertion) and neurons (education) over the life span improves function and extends quality-adjusted life years (QALYs). A challenge for research is to create a schedule for implementing interventions in optimal developmental windows. Thus, it should be possible to develop a step-care strategy analogous to others proposed for intervening in the overnutrition diathesis.[6]

Lifestyle and other remedies

Diet. More than a century of experimenting with dietary changes has not resulted in any major advances in treatment or prevention of syndromic overnutrition, despite widespread recognition of the importance of environmental factors, discoveries of molecular mechanisms, and implementation of public health measures. The reason is simply the plethora of evolutionarily conserved, redundant compensatory adaptive mechanisms resisting transient or long-term stressful undernutrition, whether imposed by natural or man-made catastrophes, physiologically comparable to dieting or restrained eating.[8,9]

The re-awakened interest in the gut microbiome, embodied in recognition of the phenomenon of dysbiosis, has the potential for revitalizing nutrition science via introduction of protective probiotics, nutraceuticals, judiciously targeted antibiotics, and fecal microbiota transplantation (FMT). Indeed, beneficial effects of FMT have been reported for metabolic syndrome and diverse neurologic disorders.[10]

Exercise. The meeting abundantly exposed the salutary preventive benefits of physical activity—the other side of energy equilibrium—for combating sarcopenia, frailty, falls, and dementia. There is ample evidence that exercise reduces inflammation, insulin resistance, and all components of metabolic syndrome, prevents some forms of cancer, improves mood, and prolongs life,[11] as does sustained weight loss.[12] However, exercise or exertion, the other lifestyle factor, has not been widely adopted for combating adiposity, inflammation, insulin resistance, and dyslipidemia, although it is true that several successful, albeit relatively short-term, community-level trials have been completed. Increasing overall energy expenditure over the lifecycle will require major educational and policy measures to combat the comfort of sedentism in developed nations.

Pharmacology and surgery. Even if evidence associating T2D with AD[13] and supporting early aggressive diagnosis and treatment of hyperinsulinemia for prevention[14] is powerful—resveratrol and other Sirtuin 1–activating compounds (STACs) with palliative potential for age-related diseases[15] have made a comeback and new promising targets have emerged[16]—it is unlikely that drug treatment can be offered on a large scale. Metabolic diabesity surgery is remarkably effective[17] but is also not a realistic option on a population level. Here, too, wide-ranging policy measures are urgently needed.

Conclusion

Achieving the goal of reestablishing eumetabolic isoenergetic energy balance will require more than top–down exhortations to "eat less" and "exercise more." It will require research focusing on the foundations of societal, self-perpetuating, and relatively short-sighted biases that similarly have evolved, owing to their survival value, only to become maladaptive during the last century. These biases, often dogmatic, are manifested in global movements such as communism, capitalism, industrialization and other cultural belief systems. The recent globalization of communication is confronting the dogma of these belief systems. The real challenge will be to define and incentivize universally beneficial behaviors, in essence changing cultures—acculturation—in different regions of the planet.

References

1. Olshansky, S.J., B.A. Carnes, R. Hershow, *et al.* 2005. Misdirection on the road to Shangri-La. *Sci. Aging Knowledge Environ.* **28:** pe15.
2. Lozano, R., M. Naghavi, K. Foreman, *et al.* 2010. Global and regional mortality from 235 causes of death for 20 age groups in 1990 and 2010: a systematic analysis for the Global Burden of Disease Study. *Lancet* **380:** 2095–2128.
3. Kral, J.G., P. Otterbeck & M. Garcia-Touza. 2010. Preventing and treating the accelerated ageing of obesity. *Maturitas* **66:** 223–230.
4. Heymsfield, S.B. & W.T. Cefalu. 2013. Does body mass index adequately convey a patient's mortality risk? *JAMA* **309:** 87–88.
5. Reiman, E.M., Y.T. Quiroz, A.S. Fleisher, *et al.* 2012. Brain imaging and fluid biomarker analysis in young adults at genetic risk for autosomal dominant Alzheimer's disease in the presenilin 1 E280A kindred: a case-control study. *Lancet Neurol.* **11:** 1048–1056.
6. Kral, J.G. 2004. Preventing and treating obesity in girls and young women to curb the epidemic. *Obes. Res.* **12:** 1539–1546.
7. Danese, A. & B.S. McEwen. 2012. Adverse childhood experiences, allostasis, allostatic load, and age-related disease. *Physiol. Behav.* **106:** 29–39.
8. Rideout, C.A., W. Linden & S.I. Barr. 2006. High cognitive dietary restraint is associated with increased cortisol excretion in postmenopausal women. *J. Gerontol. A. Biol. Sci. Med. Sci.* **61:** 628–633.
9. Kral, J.G. 2009. Psychosurgery for obesity. *Obesity Facts* **2:** 339–341.
10. Borody, T.J. & J. Campbell. 2012. Fecal microbiota transplantation: techniques, applications, and issues. *Gastroenterol. Clin. North Am.* **41:** 781–803.
11. Fortes, C., S. Mastroeni, A. Sperati, *et al.* 2013. Walking four times weekly for at least 15min is associated with longevity in a cohort of very elderly people. *Maturitas* **24:** 246–251.
12. Saczynski, J.S., S. Siggurdsson, P.V. Jonsson, *et al.* 2009. Glycemic status and brain injury in older individuals: the age gene/environment susceptibility-Reykjavik study. *Diabetes Care* 1608–1613.
13. Mayeda, E.R., M.N. Haan, A.M. Kanaya, *et al.* 2013. Type 2 Diabetes and 10-Year Risk of Dementia and Cognitive Impairment Among Older Mexican Americans. *Diabetes Care.* 2013 Mar 20. [Epub ahead of print]
14. Strittmatter, W.J. 2012. Alzheimer's disease: the new promise. *J. Clin. Invest.* **122:** 1191. Re: Talbot, K., H.Y. Wang, H. Kazi, *et al.* Demonstrated brain insulin resistance in Alzheimer's disease patients is associated with IGF-1 resistance, IRS-1 dysregulation, and cognitive decline. *J. Clin. Invest.* **122:** 1316–3138.
15. Hubbard B.P., A.P. Gomes, H. Dai, *et al.* Evidence for a common mechanism of SIRT1 regulation by allosteric activators. *Science* **339:** 1216–1219.
16. Johnson S.C., P.S. Rabinovitch & M. Kaeberlein. 2013. mTOR is a key modulator of ageing and age-related disease. *Nature* **493:** 338–345.
17. Carlsson L.M., M. Peltonen, S. Ahlin, *et al.* 2012. Bariatric surgery and prevention of type 2 diabetes in Swedish obese subjects. *N. Engl. J. Med.* **367:** 695–704.

Ann. N.Y. Acad. Sci. ISSN 0077-8923

ANNALS OF THE NEW YORK ACADEMY OF SCIENCES

Issue: Annals *Meeting Reports*

Vitamin D: beyond bone

Sylvia Christakos,[1] Martin Hewison,[2] David G. Gardner,[3] Carol L. Wagner,[4] Igor N. Sergeev,[5] Erica Rutten,[6] Anastassios G. Pittas,[7] Ricardo Boland,[8] Luigi Ferrucci,[9] and Daniel D. Bikle[10]

[1]Department of Biochemistry and Molecular Biology, UMDNJ-New Jersey Medical School, Newark, New Jersey. [2]Department of Orthopedic Surgery, The David Geffen School of Medicine at the University of California, Los Angeles, California. [3]Diabetes Center and Department of Medicine, University of California, San Francisco, California. [4]Division of Neonatology, Department of Pediatrics, Medical University of South Carolina, Charleston, South Carolina. [5]Department of Health and Nutritional Sciences, South Dakota State University, Brookings, South Dakota. [6]Ciro, Center of Expertise for Chronic Organ Failure, Horn, the Netherlands. [7]Division of Endocrinology, Diabetes and Metabolism, Tufts Medical Center, Boston, Massachusetts. [8]Departamento de Biología, Bioquímica, y Farmacia, Universidad Nacional del Sur, Bahia Blanca, Argentina. [9]Longitudinal Studies Section, National Institute on Aging, National Institutes of Health, Baltimore, Maryland. [10]Departments of Medicine and Dermatology, University of California San Francisco, and VA Medical Center, San Francisco, California

Address for correspondence: annals@nyas.org

In recent years, vitamin D has been received increased attention due to the resurgence of vitamin D deficiency and rickets in developed countries and the identification of extraskeletal effects of vitamin D, suggesting unexpected benefits of vitamin D in health and disease, beyond bone health. The possibility of extraskeletal effects of vitamin D was first noted with the discovery of the vitamin D receptor (VDR) in tissues and cells that are not involved in maintaining mineral homeostasis and bone health, including skin, placenta, pancreas, breast, prostate and colon cancer cells, and activated T cells. However, the biological significance of the expression of the VDR in different tissues is not fully understood, and the role of vitamin D in extraskeletal health has been a matter of debate. This report summarizes recent research on the roles for vitamin D in cancer, immunity and autoimmune diseases, cardiovascular and respiratory health, pregnancy, obesity, erythropoiesis, diabetes, muscle function, and aging.

Keywords: vitamin D; cancer; immunity; pregnancy; obesity; diabetes; pulmonary disease; muscle; aging; cognitive function

Introduction

Traditionally, vitamin D has been considered almost exclusively in the context of its role in calcium homeostasis. Vitamin D can either be taken in the diet, largely though fortification of dairy products, or synthesized in the skin from 7-dehydrocholesterol upon exposure to UV irradiation. Whether ingested or synthesized, vitamin D is transported to the liver, where it is hydroxylated at position 25 to form 25-hydroxyvitamin D ($25(OH)D_2$ or $25(OH)D_3$), the major circulating form of vitamin D. Next, $25(OH)D_3$ is transported to the kidney, where it is hydroxylated at position 1 by the enzyme CYP27B1 to form 1,25-dihydroxyvitamin D ($1,25(OH)_2D_3$), the most active form of vitamin D, which is then transported to target tissues, where it functions like a steroid, binding to the vitamin D receptor (VDR).

The VDR heterodimerizes with the retinoid X receptor (RXR), and the VDR/RXR complex binds to VDR-responsive elements found in or around target genes and, in association with various co-activators, results in the transcription of target genes. When there is a need to increase blood calcium levels (e.g., during growth or pregnancy), $1,25(OH)_2D_3$ acts in the intestine to increase calcium absorption. If this increased intestinal absorption is insufficient to restore normal calcium levels, $1,25(OH)_2D_3$ works in concert with the parathyroid hormone (PTH) in the kidney to promote calcium reabsorption from the distal tube, and in the skeletal system to release calcium from bones.

While it was long held that vitamin D acted only at the intestine, kidney, and skeleton, and that its function was limited to calcium homeostasis, the possibility of extraskeletal effects has been

doi: 10.1111/nyas.12129

considered for decades as a result of the discovery of the VDR in tissues that have no involvement in calcium homeostasis (e.g., skin, placenta, pancreas, breast, prostate and colon cancer cells, and activated T cells). Discovery of the VDR and CYP27B1 in these tissues led to exploration of the roles and mechanisms of vitamin D function in each. On September 21, 2012, the Abbott Nutrition Health Institute and the Sackler Institute for Nutrition Science at the New York Academy of Sciences sponsored a conference, "Vitamin D: Beyond Bone," that gathered researchers investigating vitamin D in a wide range of tissues and diseases.

Nonclassical effects of vitamin D

Vitamin D: beyond bone

Sylvia Christakos (UMDNJ-New Jersey Medical School) began the meeting by summarizing the metabolism and activity of vitamin D and introducing the emerging awareness of roles for vitamin D beyond calcium homeostasis and bone metabolism. Evidence in the laboratory, including the use of animal models, indicates that $1,25(OH)_2D_3$ generates a number of extraskeletal effects, including inhibition of cancer progression, effects on the cardiovascular system and skin, modulation of innate immunity with subsequent killing of bacteria, and inhibition of certain autoimmune diseases (reviewed in Ref. 1). For example, in rats treated with the chemical carcinogen *N*-methyl *N*-nitrosourea (NMU), $1,25(OH)_2D_3$ inhibits the progression of breast tumors. In addition, when $1,25(OH)_2D_3$ is given prior to NMU, tumor incidence is prevented or reduced.[1] $1,25(OH)_2D_3$ also reduces the incidence and severity of prostate neoplasia in a mouse model (Ndx 3.1; *Pten* mutant mouse), and $1,25(OH)_2D_3$ has tumor inhibitory activity in a mouse model of colorectal adenoma (Apc^{min}). In order to determine mechanisms involved in inhibition of breast tumor growth, Christakos' lab showed that C/EBPα, a transcription factor that has been shown to play a critical role in growth arrest of other cell types, is induced by $1,25(OH)_2D_3$ in MCF-7 human breast cancer cells.[2] C/EBPα was found to induce transcription of the vitamin D receptor in MCF-7 cells.[2] Since the levels of the VDR correlate with the antiproliferative effects of $1,25(OH)_2D_3$, and since it has been suggested that C/EBPα can be considered a potential tumor suppressor, these findings suggest mechanisms whereby $1,25(OH)_2D_3$ may

act to inhibit growth of breast cancer cells. These findings also identify C/EBPα as a $1,25(OH)_2D_3$ target in breast cancer cells and provide evidence for C/EBPα as a candidate for breast cancer treatment.[2]

With regard to autoimmune diseases, $1,25(OH)_2D_3$ has been shown to suppress type 1 diabetes in the non-obese diabetic (NOD) mouse model, to suppress experimental autoimmune encephalomyelitis (EAE) (a mouse model of multiple sclerosis (MS)), and to suppress mouse models of inflammatory bowel disease and systemic lupus erythematosus.[1] Recent studies from Christakos' lab have shown that inhibition of EAE is associated with inhibition of interleukin (IL)-17, a cytokine that plays a critical role in numerous inflammatory conditions and autoimmune diseases including MS. The mechanism of $1,25(OH)_2D_3$ suppression of IL-17 was found to be transcriptional and to involve blocking of nuclear factor for activated T cells (NFAT, which is important for T cell receptor–mediated transcriptional regulation of IL-17), recruitment of histone deacetylase to the IL-17 promoter, and sequestration of Runt-related transcription factor 1 (Runx1) by the VDR.[3] $1,25(OH)_2D_3$ was also found to have a direct effect on the induction of Foxp3, a transcription factor that associates with NFAT and Runx1 for transcriptional repression.[3] These results describe novel mechanisms and new concepts with regard to vitamin D and the immune system and suggest therapeutic targets for the control of autoimmune diseases.

Unlike the association between vitamin D deficiency and rickets, causal links between vitamin D deficiency and specific extraskeletal diseases have yet to be identified. However, the evidence in the laboratory of beneficial effects of $1,25(OH)_2D_3$ beyond bone is compelling (summarized in Fig. 1). Findings in animal models may suggest mechanisms involving similar pathways in humans that could lead to the identification of new therapies.

Vitamin D in immune function and disease prevention

Martin Hewison (the David Geffen School of Medicine, University of California) detailed one of the most prominent of the so-called nonclassical effects of vitamin D: its ability to act as a potent modulator of human immune responses. Evidence for this initially stemmed from two observations.

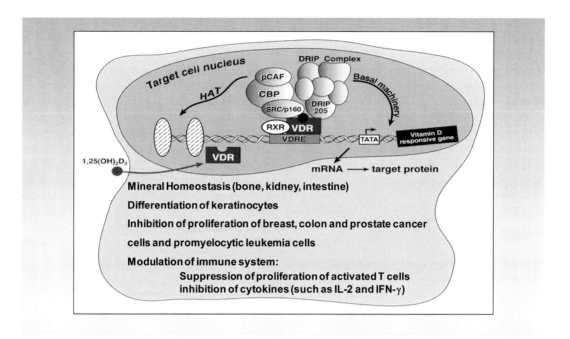

Figure 1. Genomic mechanism of vitamin D action. Mechanism of action of 1,25(OH)$_2$D$_3$ in target cells. The VDR heterodimerizes with the RXR. After interaction with the VDRE (vitamin D response element), transcription proceeds through the interaction of the VDR with coactivators and with the transcription machinery. The histone acetyltransferase (HAT) activity–containing complex (SRC/p160 and CBP), the DRIP complex, and additional coactivators not shown (including specific methyltransferases) are recruited by liganded VDR. 1,25(OH)$_2$D$_3$ is known to maintain calcium homeostasis and to affect numerous other cell types. Effects on other cell systems, including modulation of the immune system and inhibition of proliferation of cancer cells, are discussed. With permission from Christakos.[60]

First, many cells from both the innate and adaptive immune systems express the VDR. Second, antigen cells from the innate immune system, such as macrophages or dendritic cells (DCs), also express the vitamin D activation enzyme 1α-hydroxylase, also known as CYP27B1. As such, these cells are able to convert precursor 25(OH)D$_3$, the major circulating form of vitamin D, to active 1,25(OH)$_2$D$_3$ that can then induce responses in the cells by binding to their VDRs and promoting transcriptional regulation.

This localized intracrine mechanism appears to be central to two key features of immune function: innate antibacterial activity and the presentation of antigen to cells from the adaptive immune system such as T lymphocytes (T cells). In macrophages and monocytes, cellular sensing of pathogens, such as *Mycobacterium tuberculosis*, by pattern recognition receptors, such as Toll-like receptors, enhances expression of CYP27B1 and the VDR, thereby increasing intracrine synthesis and activity of 1,25(OH)$_2$D$_3$.[4] The resulting effects, including enhanced expression of the antibacterial

proteins cathelicidin and β-defensin 2, and enhanced formation of autophagosomes, facilitate improved bacterial killing in a vitamin D–dependent fashion. Vitamin D insufficiency or deficiency may therefore impair innate immune responses and predispose individuals to infection, and low serum concentrations of 25(OH)D$_3$ have been linked to infectious diseases such as tuberculosis. However, it is important to recognize that similar antibacterial responses to vitamin D have also been reported for other cell types within tissues, such as the skin, lungs, GI tract, and placenta, that may broadly be termed *barrier sites*. Thus, it is possible that vitamin D deficiency has a generalized detrimental impact on antibacterial responses and may therefore play a role in many types of infectious diseases.[5] An important note by Hewison is that current data suggest that this particular immune response to vitamin D is restricted to primates, underlining a potential role for vitamin D in the evolution of primate immunity, but also limiting the scope of further *in vivo* studies to explore this activity.

The immunomodulatory effects of vitamin D also involve the adaptive immune system. Intracrine synthesis of $1,25(OH)_2D_3$ by DCs decreases their maturation, thereby suppressing antigen presentation and decreasing T cell proliferation.[6] Recent studies have shown that intracrine expression and activity of CYP27B1 by DCs is particularly effective in supporting the generation of tolerogenic regulatory T (T_{reg}) cells, while simultaneously suppressing inflammatory IL-17–expressing T cells (Th17 cells).[7] In this way, localized metabolism of $25(OH)D_3$ may play a pivotal role in prevention and/or treatment of autoimmune diseases, such as multiple sclerosis, rheumatoid arthritis, and inflammatory bowel disease. As with antibacterial responses to infection, it has been proposed that adaptive immune responses may be compromised in patients with low serum concentrations of $25(OH)D_3$, and association studies have shown links between autoimmune disease and vitamin D deficiency. The extent to which vitamin D deficiency is a cause of immune disease has yet to be demonstrated and will require new prospective clinical trials of vitamin D supplementation. Hewison stated that it will be important to define whether enhanced vitamin D–mediated immunity prevents immune disease or whether it can be used as therapy for these diseases. Moreover, it is possible that some diseases, such as leprosy and HIV infection, may be associated with dysregulation of the core vitamin D intracrine immune mechanism, and may therefore require higher circulating levels of $25(OH)D_3$ to reestablish normal immune responses. These and other facets of vitamin D and immune function will be important research challenges for the vitamin D community.

Vitamin D in the heart and vascular system

David G. Gardner (University of California, San Francisco) expanded the discussion of nonclassical effects of vitamin D to include evidence of a significant role for vitamin D in the cardiovascular system. While the heart and vasculature have only recently been identified as potential targets of vitamin D action, a growing body of evidence suggests that the activated VDR plays an important role in regulating cardiovascular function.[8] First, the VDR and the ligand-generating 25-hydroxyvitamin D_3 1-α-hydroxylase have been shown to be present in cardiac myocytes, cardiac fibroblasts, vascular

smooth muscle cells, and vascular endothelial cells. Second, vitamin D deficiency in rodents has been shown to elicit increases in blood pressure and cardiac hypertrophy. Third, $1,25(OH)_2D_3$ and its analogs have been shown to reverse agonist-induced myocyte hypertrophy *in vitro* and endogenous cardiac hypertrophy in the Dahl S rat, the spontaneous hypertensive rat (SHR), the spontaneous hypertensive heart failure–prone rat (SHR-HF), and the 5/6 nephrectomy model of chronic renal failure, *in vivo*. Fourth, a variety of cardiovascular disorders including congestive heart failure, coronary artery disease, and peripheral vascular disease have been linked to reduced circulating levels of $25(OH)D_3$. Fifth, and perhaps most important, a mouse with a complete deletion of the VDR gene demonstrates both hyperreninemic hypertension and cardiac hypertrophy.[9]

Gardner and colleagues examined the ability of paricalcitol, a bioactive analog of $1,25(OH)_2D_3$, to prevent cardiac hypertrophy in rats infused with moderate doses of angiotensin II (800 ng/kg/min) over a 14-day period. Infusion of angiotensin II led to increased blood pressure, increased myocyte hypertrophy, increased expression of the hypertrophy-sensitive fetal gene program (i.e., atrial natriuretic peptide, B-type natriuretic peptide, and alpha skeletal actin gene expression), and increased cardiac interstitial fibrosis with augmented procollagen 1 and 3 expression. In each case, coadministration of paricalcitol (intraperitoneal injection of 300 ng/kg every 48 h), resulted in partial reversal of the angiotensin II effects.[10]

In an effort to define the role of the VDR in the myocardium, Gardner and colleagues created a mouse with a selective deletion of the fourth exon of the murine *Vdr* gene in cardiac myocytes. This mouse displayed an increase in myocyte size, left ventricular weight/body weight, and hypertrophy-dependent fetal gene expression compared with control littermates. There was no demonstrable increase in interstitial fibrosis. The mouse also demonstrated a reduction in end-diastolic and end-systolic volume by echocardiography and increased expression of modulatory calcineurin inhibitory protein 1 (MCIP1), a direct downstream target of calcineurin and NFAT that has been linked to the development of cardiac hypertrophy. Gardner's group showed that isoproterenol treatment of neonatal rat cardiac myocytes *in vitro* resulted in myocyte hypertrophy and increased MCIP1 expression. Coadministration of $1,25(OH)_2D_3$ resulted in a dose-dependent reduction in MCIP1 expression.[11]

Vitamin D and the VDR have also been implicated in the support of normal endothelial function. Endothelial dysfunction, regarded by many as a harbinger of cardiovascular disease, has been linked to a number of disorders that are also associated with low circulating levels of $25(OH)D_3$. In some cases, treatment of these disorders with exogenous vitamin D or vitamin D metabolites has been shown to restore endothelial function. To establish a link between the VDR and endothelial function, Gardner and colleagues created a mouse with selective deletion of the *Vdr* gene in vascular endothelial cells. This mouse displays many of the phenotypic features normally associated with endothelial dysfunction.

Collectively, these data demonstrate that the vitamin D system—specifically, the ligand-bound VDR—plays an important role in the maintenance of cardiovascular health and suggest that vitamin D repletion may have important public health implications in controlling the prevalence of, and morbidity associated with, cardiovascular disease. Of equivalent importance, therapy with high potency vitamin D metabolites or their analogs may subserve an important role in the management of these diseases.

Vitamin D during pregnancy and lactation

Vitamin D as a preprohormone has effects that extend beyond calcium metabolism and homeostasis throughout life but which are most vivid during pregnancy when vitamin D metabolism is disengaged from its usual constraints.[12] At no other time during the lifecycle is $25(OH)D_3$ linked directly to $1,25(OH)_2D_3$ production, the latter rising more than 2.5 times above nonpregnant levels.[12] This elevation of $1,25(OH)_2D_3$ likely serves the purpose of immune regulation, though there have been only a few studies that have evaluated the effect of vitamin D status on immune function during pregnancy to support this hypothesis. In addition, few randomized controlled trials have been conducted in pregnant women to determine optimal vitamin D status.[13–15]

Carol L. Wagner (Medical University of South Carolina) discussed two recent studies performed by her group in collaboration with Bruce Hollis's group—the National Institute of Child Health and Human Development (NICHD) Vitamin D Pregnancy Trial[12] and the Thrasher Research Fund Vitamin D Community-Based Supplementation Trial[16]—that indicate that 400 IU vitamin D, the amount found in most prenatal vitamins, is inadequate and that 4000 IU/day is necessary to optimize $1,25(OH)_2D_3$ production, achieved when total circulating $25(OH)D_3$ is at least 40 ng/mL (Fig. 1).[12] When analyzed by maternal $25(OH)D_3$ concentration at various time points—i.e., baseline, by trimester, throughout pregnancy as mean, median, and area under the curve, or at delivery— the group taking 4000 IU/day attained optimal $1,25(OH)_2D_3$ conversion from $25(OH)D_3$ throughout pregnancy at considerably higher rates than groups taking lower doses. Both studies provide evidence that 4000 IU/day is not only safe but is associated with fewer adverse events of pregnancy when taken together as comorbidities of pregnancy, compared with the lower dose groups.

Comorbidities of pregnancy appear to be directly linked to lower $25(OH)D_3$, and, conversely, increasing $25(OH)D_3$ appears to afford protection to the mother and developing fetus, with lower risk of preterm labor, preterm birth, or infection for each 10 ng/mL increase in total circulating $25(OH)D_3$ concentration.[16] Such findings suggest a simple approach to reduce morbidities during pregnancy. Yet, before such findings are fully embraced, the mechanism(s) of action of vitamin D and its protective effect during pregnancy must be better delineated. Additional research that addresses the impact of vitamin D on immune homeostasis during pregnancy and its effects during early infancy and later childhood becomes essential.

These findings extend to lactation and early infancy. It was thought for decades that human milk was minimally sufficient in vitamin D;[17] yet, preliminary results of the recently completed NICHD vitamin D supplementation trial during lactation show that when a mother is vitamin D replete, with an intake of 6400 IU vitamin D_3/day, breast milk is replete and a breastfeeding infant has excellent vitamin D status without infant supplementation that is comparable to combined maternal and infant supplementation of 400 IU vitamin D_3/day.[18] The implications of safely achieving vitamin D sufficiency for both mother and breastfeeding infant solely through maternal supplementation is just beginning to be understood and will likely challenge researchers for decades to come.

Vitamin D and obesity

Igor N. Sergeev (South Dakota State University) discussed the possibility of integrating vitamin D supplementation with current strategies in the prevention and treatment of obesity. The induction of adipocyte death through apoptosis is emerging as a promising strategy for addressing obesity.[19,20] Increased adipose tissue mass is the result of both hypertrophy (an increase in adipocyte size) and hyperplasia (an increase in adipocyte number), and once adipocytes achieve a maximum size, further increase in adipose tissue mass involves an increase in adipocyte number. Thus, weight loss can result from not only a decrease in adipocyte size but also adipocyte number. Even a small increase in the amount of adipocyte apoptosis will prevent excessive accumulation of adipose tissue and can result in a significant loss of adipose tissue mass over time. Therefore, removal of adipocytes through apoptosis reduces body fat and can help in long-lasting maintenance of weight loss.

The effects of the hormonal form of vitamin D, $1,25(OH)_2D_3$, on apoptotic cell death are mediated via multiple signaling pathways that involve common regulators and effectors converging on cellular Ca^{2+}.[19,21] Sergeev and colleagues have shown that the critical characteristic of the apoptotic Ca^{2+} signal is a sustained increase in the concentration of intracellular Ca^{2+}, reaching elevated but not cytotoxic levels.[22] However, $1,25(OH)_2D_3$-regulated, Ca^{2+}-dependent apoptotic molecular targets have not been identified in adipose tissue. Thus, Sergeev and colleagues have been investigating the mechanism by which $1,25(OH)_2D_3$ regulates apoptosis in adipocytes. Their results have demonstrated that $1,25(OH)_2D_3$ induces, in a concentration- and time-dependent fashion, a sustained, prolonged increase in concentration of intracellular Ca^{2+} (the apoptotic Ca^{2+} signal) in mature mouse 3T3-L1 adipocytes.[23] The $1,25(OH)_2D_3$-induced increase in cellular Ca^{2+} is associated with activation of Ca^{2+}-dependent μ-calpain and Ca^{2+}/calpain-dependent caspase-12.[23] Activation of these proteases is sufficient to effect morphological and biochemical changes consistent with apoptosis. The $1,25(OH)_2D_3$-induced

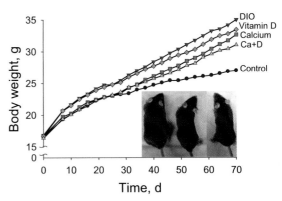

Figure 2. Body-weight gain in mice fed high-fat diets with increased levels of Ca and vitamin D. Weight matched mice were randomly assigned to the experimental high-fat (HF) diets containing 60% energy as fat (the normal control diet contained 10% energy as fat). The treatment groups on HF diet were high-calcium (Ca diet, 1.2% Ca), high-vitamin D_3 (D diet; vitamin D_3 intake 10 times higher than the recommended level of 1000 IU/kg) and high-Ca plus high-D_3 (Ca + D diet). Data are means for each group and time point; $n = 7$–8 per group from week 1 to week 10. The insert shows (from left to right) mice from the DIO, control, and Ca + D groups at week 10. Data are unpublished and from work in progress.

increase in cellular Ca^{2+} is also associated with reduced lipid accumulation in mature adipocytes.[23] In prelimary but ongoing studies, a murine diet-induced obesity (DIO) model is being used to evaluate the role of vitamin D in adiposity. DIO mice (C57BL/6J) fed a high-vitamin D_3 diet, in particular a high-vitamin D_3 plus high-calcium diet, show decreased body and fat weight gain (Fig. 2; unpublished data) and improved markers of adiposity and vitamin D status (plasma concentrations of glucose, insulin, adiponectin, $25(OH)D_3$, $1,25(OH)_2D_3$, and PTH), but an increased plasma Ca^{2+}. High vitamin D_3 and calcium intake is associated with activation of the Ca^{2+}-dependent apoptotic proteases calpain and caspase-12 in adipose tissue of DIO mice. These preliminary findings suggest that the $1,25(OH)_2D_3$-induced cellular Ca^{2+} signal can act as an apoptotic initiator that directly recruits Ca^{2+}-dependent apoptotic effectors capable of executing apoptosis in adipose tissue. The preliminary results also suggest that high vitamin D and calcium intake decreases body and fat weight gain in diet-induced obesity, and that the potential mechanism of these effects involves activation of a Ca^{2+}-mediated apoptotic pathway in adipose tissue (specifically, normalization of the activities of apoptotic proteases

Ca^{2+}-dependent calpain and Ca^{2+}/calpain-dependent caspase-12 that are reduced in obesity). Targeting Ca^{2+} signaling and the vitamin D/Ca^{2+}–dependent calpains and caspases in adipocytes with vitamin D supplementation may therefore be an effective and affordable approach for chemoprevention and treatment of obesity. Additional studies are warranted to evaluate this approach from a safety point of view and to identify the optimal levels of calcium and vitamin D intakes in obesity.

Vitamin D and lung function in patients with chronic obstructive pulmonary disease

Erica Rutten (Ciro+, Centre of Expertise for Chronic Organ Failure, Netherlands) discussed her work investigating the association between plasma vitamin D levels and lung function in chronic obstructive pulmonary disease (COPD) patients. COPD is characterized by a persistent and, usually, progressive airflow limitation and is associated with an enhanced local inflammatory response. By 2030, COPD is predicted to become the fourth leading cause of death worldwide, and therefore accounts for about 8% of total deaths. Today, in addition to lung function impairment, COPD is recognized as a systemic disease with multiple comorbidities, such as osteoporosis, cardiovascular disease, renal impairment, and psychological disorders. In relation to the systemic pathology of COPD the role of vitamin D has generated interest, as vitamin D deficiency is often present in patients with COPD,[24] dependent on the severity of COPD.[25] Indeed, the prevalence of vitamin D deficiency, defined as plasma 25(OH) D_3 concentrations below 20 ng/ml (50 nmol/L), was 58% in a cross-sectional study of 151 COPD patients entering pulmonary rehabilitation during summer.[26] In the Bergen COPD Cohort Study, the prevalence of vitamin D deficiency was indeed higher in the COPD patients than in the control subjects, independent of season, age, smoking, comorbidity, and body mass index.[27] Several potential explanations for the increased risk of vitamin D deficiency in patients with COPD include poor diet, less outdoor activity and thus less exposure to sunlight, accelerated skin aging due to smoking, renal dysfunction, and treatment with corticosteroids. Vitamin D is related to skeletal health in patients with COPD, and vitamin D deficiency is associated with low bone mineral density.[26, 28] Additionally, low vitamin D has been related to decreased exercise performance[26, 29] and

exacerbation rate[30] in COPD patients. And while vitamin D concentration has been related to lung function in the general population, the relation between lung function and vitamin D concentration has yet to be investigated in COPD patients only.

Rutten and colleagues performed a secondary analysis of the publication by Romme *et al.*[26] to investigate whether there is an independent association between lung function parameters and plasma vitamin D levels in a group of 151 COPD patients admitted for pulmonary rehabilitation at Ciro+. Plasma 25(OH)D_3 levels, body mass index (BMI; weight in kg/(height in m)^2), and lung function parameters (forced expiratory volume in 1s (FEV1), forced vital capacity (FVC), and diffusing capacity of the lung (DLCO)) were measured. The study group comprised 57% males, with mean age 65 ± 9 years, mean FEV1 47.2 ± 17.9% predicted, mean FVC 94.6 ± 20.6% predicted, and mean DLCO 53.2 ± 19.7% predicted; 58% of the patients were vitamin D deficient (vitamin D concentration <20 ng/ml). Pearson correlation coefficient showed significant correlations between plasma vitamin D concentration and FEV1 ($r = 0.31$, $P < 0.01$), FVC ($r = 0.29$, $P < 0.01$) and DLCO ($r = 0.20$, $P = 0.01$). After stratification for vitamin D deficiency, the correlations persisted only in the patients with vitamin D deficiency: FEV1 ($r = 0.37$, $P < 0.01$; Fig. 3, data are from an unpublished study), FVC ($r = 0.25$, $P < 0.01$), and DLCO ($r = 0.32$,

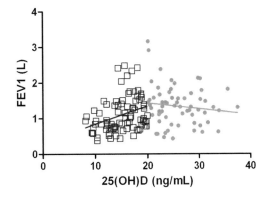

□ Patients with vitamin D deficiency: r = 0.37, p < 0.01
● Patients without vitamin D deficiency: r = -0.08, p = 0.53

Figure 3. Lung function and plasma vitamin D levels in a group of 151 COPD patients. FEV1, forced expiratory volume in 1 s. These data are from an unpublished secondary analysis by Ruttan and colleagues.

$P = 0.02$). Multivariate regression analyses revealed that, after correction for age, gender, and BMI, vitamin D concentration remained independently associated with FEV1 ($\beta = 0.26$, $P < 0.01$) and DLCO ($\beta = 0.25$, $P < 0.01$), and there was a trend for FVC ($\beta = 0.15$, $P < 0.08$) in the subjects with vitamin D deficiency. From their unpublished secondary analysis, Rutten and colleagues conclude that there is evidence suggesting that vitamin D plays a role in the lung pathology of patients with COPD, which requires further investigation.

Impact of cholecalciferol repletion on erythropoietin requirements in hemodialysis patients

Lily Li (Cleveland Clinic Lerner College of Medicine of Case Western Reserve University and Mount Sinai School of Medicine) presented pilot data from a randomized control trial investigating the impact of cholecalciferol repletion on erythropoietin requirements in vitamin D–deficient hemodialysis patients. Li mentioned that immune cells contain the machinery needed to convert $25(OH)D_3$ to $1,25(OH)_2D_3$, and that there is evidence that this local production of $1,25(OH)_2D_3$ has immunomodulatory properties. In a setting of decreased $25(OH)D_3$, the absence of $1,25(OH)_2D_3$ production may enhance local inflammation.

Among patients with end-stage renal disease (ESRD) on hemodialysis, $25(OH)D_3$ deficiency is common, reported at greater than 80%.[32] ESRD patients are commonly treated with $1,25(OH)_2D_3$ for prevention of secondary hyperparathyroidism and bone loss. However, $25(OH)D_3$ is rarely repleted within this population, and the impact of $25(OH)D_3$ deficiency remains obscure. Anemia is also associated with ERSD and is the result of reduced erythropoietin production due to kidney disease as well as iron deficiency and inflammation-mediated abnormalities in iron metabolism. In this population, anemia not only causes impaired quality of life but also left ventricular hypertrophy, myocardial infarction, and increased mortality. Patients with ESRD-associated anemia are commonly treated with erythropoiesis-stimulating agents (ESA), which are very effective in the majority of the population. However, a subset of these patients develop erythropoietin hyporesponsiveness, which is thought to be caused by inadequate iron mobilization in the setting of inflammation, medi-

ated in part by hepcidin, a small molecule produced by hepatocytes that is known to inhibit macrophage iron release and intestinal iron absorption. Since hepcidin production appears to be induced by IL-6, and $25(OH)D_3$ has been shown to suppress IL-6 production in monocytes,[33] Li's group developed the hypothesis that in hemodialysis patients deficiency of $25(OH)D_3$, which is required for local production of $1,25(OH)_2D_3$ in immune cells, leads to dysregulation of innate immunity and inflammation (e.g., enhanced IL-6 production), which alters iron metabolism and contributes to erythropoietin resistance.

They have initiated a randomized controlled trial of $25(OH)D_3$ repletion in vitamin D–deficient hemodialysis patients to evaluate efficacy and safety and the effects on inflammation and ESA requirements. Patients are given either no repletion (standard of care) or 50,000 IU/week of cholecalciferol for six weeks, or until $25(OH)D_3$ levels exceeded 35 ng/mL, and 10,000 IU/week afterwards. Their preliminary results show that $25(OH)D_3$ repletion was both safe and effective, with $25(OH)D_3$ levels increasing significantly in the treatment arm and no patients experiencing hypercalcemia or other adverse effects. In addition, patients in the treatment arm exhibit a decrease in ESA requirement at six months. Li and colleagues have isolated monocytes from patients and examined expression of a number of genes related to vitamin D metabolism and immune responses. Their observations include a decrease in IL-6 expression following $25(OH)D_3$ repletion at six months. These preliminary data indicate that $25(OH)D_3$ repletion treatment is safe, effective, and may result in lower ESA requirements, a result with significant economic implications.

Translating epidemiological data into policy and clinical applications

Vitamin D for diabetes

Anastassios G. Pittas (Tufts Medical Center) highlighted the potentially important role of vitamin D in a variety of non-skeletal medical conditions, including type 2 diabetes. To systemically appraise the available evidence on the role of vitamin D in diabetes, Pittas reviewed the original Bradford Hill criteria, which have been subsequently modified to include three general categories (mechanistic studies, direct evidence, and parallel evidence),[34] and then applied these criteria to the available evidence.

From a biological perspective, the hypothesis that vitamin D may be a determinant of diabetes risk is plausible, as both impaired insulin secretion and action have been reported with vitamin D insufficiency. The effect of vitamin D may be direct, as supported by the expression of the VDR and the local production of $1,25(OH)_2D_3$ in pancreatic β cells, or indirect via its role in regulating calcium homeostasis and calcium flux through cell membranes.

In humans, the role of vitamin D in type 2 diabetes is suggested by a large number of cross-sectional studies that have consistently found an inverse association between vitamin D status and prevalent hyperglycemia, including a reported seasonal variation in the control of glycemia in patients with diabetes, being worse in the winter when hypovitaminosis D is more prevalent.[35] In the majority of longitudinal observational studies, lower vitamin D status is also associated with increased risk of incident type 2 diabetes. In the most recent systematic review of observational studies, Song *et al.*[36] reported a 38% relative risk reduction in incident diabetes among individuals with the highest versus the lowest category of blood $25(OH)D_3$ concentration. In dose-response analyses, the risk of type 2 diabetes was reduced by 4% for each 4 ng/mL increment in blood $25(OH)D_3$ concentration.

However, Pittas cautioned that one should resist making inferences on the basis of observational studies because of the possibility of confounding issues, which may be especially true when assessing vitamin D status, because while $25(OH)D_3$ concentration is an excellent marker of good health many of the variables that contribute to $25(OH)D_3$ concentration (e.g., skin pigmentation, physical inactivity, aging, unhealthy dietary patterns) are also risk factors for the development of diabetes. Therefore, evidence from intervention studies—which is the most direct evidence—is critically important before one can conclude that vitamin D has a role in the prevention or treatment of diabetes. The results from small clinical trials and *post hoc* analyses of larger trials on the effect of vitamin D supplementation on diabetes-related parameters have been inconclusive, although vitamin D appears to have beneficial effects in persons at risk for diabetes.[36–39] However, no firm conclusions can be drawn from the available intervention studies because of several limitations:

(1) nearly all studies were underpowered; (2) the minority of studies were designed specifically for glycemic outcomes; (3) most studies did not report use of diabetes medications at baseline or during the study; (4) some trials used large infrequent doses of vitamin D, which may be metabolized differently compared to daily doses and may provide either no benefit or result in an unfavorable benefit/risk ratio.

Pittas concluded that although the observation studies strongly suggest an important role for vitamin D in type 2 diabetes, and such a role is biologically plausible, there is lack of evidence from intervention studies to support the contention that type 2 diabetes can be improved or prevented by raising $25(OH)D_3$ concentration. On numerous earlier occasions, encouraging findings from observational studies were not confirmed by well-designed clinical trials (e.g., hormone replacement therapy, β-carotene, vitamins C and E, and folic acid) and prevailing clinical practice was overturned. Therefore, confirmation of a potential beneficial effect of vitamin D in type 2 diabetes is needed in trials conducted in well-defined populations (e.g., pre-diabetes and early diabetes) specifically designed to test the compelling but yet unproven hypothesis that vitamin D status is a direct contributor to the pathogenesis of type 2 diabetes.

Molecular aspects of the role of vitamin D in muscle

Ricardo Boland (Universidad Nacional del Sur, Argentina) discussed molecular aspects of the role of vitamin D in skeletal muscle health and function. This is a topic of clinical significance because myopathy characterized by proximal muscle weakness and atrophy is a common symptom in vitamin D deficiency states.[40] The loss of muscular strength and a reduction in muscle mass increases the risk of falls and leads to fractures. It has been reported that administration of the hormonal vitamin D metabolite $1,25(OH)_2D_3$ decreases the number of falls by improving muscle strength in elderly subjects.[41] Consistent with this observation, there is evidence that $1,25(OH)_2D_3$ regulates muscle growth and development and contractility.[40] Moreover, the presence of the VDR in avian, murine, and human skeletal muscle has been demonstrated.[40,42]

As in other target cells, $1,25(OH)_2D_3$ elicits long-term and short-term responses in skeletal muscle that involve genomic and non-genomic modes of actions, respectively. In the first, more classical mechanism, the hormone stimulates muscle cell proliferation and differentiation through nuclear VDR-mediated gene transcription, expressed by increased myoblast DNA synthesis followed by the induction of muscle specific myosin- and calcium-binding proteins. The non-genomic effects of $1,25(OH)_2D_3$ are involved in the fast regulation of the calcium messenger system and growth-related signal transduction pathways in skeletal muscle cells. The hormone interacts with a membrane receptor, which leads to stimulation of adenylyl cyclase and phospholipases C, D, and A2 and activation of MAPK cascades.[43] Boland pointed out that there is a wealth of evidence indicating that $1,25(OH)_2D_3$ mainly acts in muscle through these membrane-initiated events and that the VDR plays a role in the activation of intracellular signaling pathways.[44]

Boland then focused his presentation on the signal transduction mechanisms of $1,25(OH)_2D_3$ in skeletal muscle. $1,25(OH)_2D_3$-induced transmembrane activation of adenylyl cyclase/cAMP/PKA and PLC/DAG plus IP_3/PKC regulates Ca^{2+} influx mediated by voltage-dependent Ca^{2+} channels (VDDC).[43] Boland's group showed that, in muscle cells, the hormone also modulates Ca^{2+} influx through SOC channels that are activated by the PLC-mediated Ca^{2+} release from intracellular stores via IP_3—a process known as capacitative calcium entry (CCE).[45] Boland discussed the finding that $1,25(OH)_2D_3$ rapidly induces reverse translocation of the VDR from the nucleus to the plasma membranes.[46] In accordance with this observation, a complex is formed between the VDR and TRCP3, an integral protein of capacitative Ca^{2+} entry, suggesting an association between both proteins and a functional role of the VDR in $1,25(OH)_2D_3$ activation of CCE. Further supporting this interpretation, transfection with a VDR antisense oligonucleotide inhibited Ca^{2+} influx through SOC channels. In some vertebrate systems, as in invertebrate photoreceptor cells, TRP proteins have been shown to interact with the PDZ domain–containing INAD proteins, which serve as acceptors for a molecule with a regulatory function, which Boland hypothesized could be the VDR. Of additional mechanistic importance, when muscle cells are treated with $1,25(OH)_2D_3$ in the presence of a specific anti-INAD antibody or transfected with an anti-INAD

antisense oligonucleotide, the hormone-dependent CCE was almost totally suppressed.[43,45]

These data enabled Boland and colleagues to propose a non-genomic mechanism by which $1,25(OH)_2D_3$ regulates intracellular Ca^{2+} levels in skeletal muscle cells: by acting on voltage-dependent channels via protein kinase–mediated phosphorylation and on SOC/TRCP3 channels through VDR-joining membrane supramolecular complexes. Alterations in these mechanisms during vitamin D deficiency states may account for skeletal muscle weakness, in view of the key role that Ca^{2+} plays in the regulation of contractility.

In addition, $1,25(OH)_2D_3$ modulates growth-related signal transduction pathways in skeletal muscle cells. It has been established that the hormone activates MAPKs (ERK1/2) and Akt. Of note, the formation of complexes between the VDR and Src is involved in the activation of these pathways by the hormone.[44,45] Experimental data demonstrates that $1,25(OH)_2D_3$ induces the translocation of activated ERK to the nucleus, where it phosphorylates the transcription factors Elk and CREB, which coregulate the expression of genes that mediate mitogenic effects of the hormone.[43] Akt is activated by $1,25(OH)_2D_3$ through PI3K, as the inhibitors Ly294002 and wortmannin block the increase in its phosphorylation in response to the hormone. Moreover, Akt mediates $1,25(OH)_2$ D_3-induced differentiation of skeletal muscle myoblasts to myotubes. With regard to the mechanism by which the hormone initiates its action at the muscle cell plasma membrane, preliminary work using MβCD and siRNA technology has shown that intact caveolae and caveolin-1 expression are involved in ERK1/2 and Akt activation by $1,25(OH)_2D_3$ (C. Buitrago & R. Boland, submitted). Boland concluded that the information described contributes to the elucidation of the mechanisms involved in the regulation of contractility and myogenesis by $1,25(OH)_2D_3$.

Vitamin D and physical and cognitive function in older persons

Luigi Ferrucci (National Institute on Aging) discussed recent evidence linking vitamin D deficiency to the pathogenesis and clinical progression of chronic diseases highly prevalent in older persons. In large population-based studies, low levels of vitamin D predict loss of mobility and disability in basic activities of daily living, independent of age and other potential confounders.[47] Based on these findings, some gerontologists claim that vitamin D has antiaging effects. Indeed, beyond its roles in bone health and calcium homeostasis, vitamin D affects many other physiological domains that are critically modified by aging. Unfortunately, the question of whether vitamin D prevents or slows aging cannot be easily tested. In spite of the intuitive nature of aging as a phenomenon, scientists have been unable to agree on good measures of biological aging because of great interindividual variability. Specific phenotypic changes occur in almost all aging individuals: changes in body composition, imbalance between production and utilization of energy, loss of redundancy of the homeostatic network, and loss of neurons and neuronal plasticity. Shifting attention from aging to the aging phenotype, the question of whether vitamin D affects major aging phenotypes becomes more approachable.

Data from the literature can generate some answers. Major changes in body composition that occur with aging include a decline in lean body mass (mostly muscle) and increase in fat mass. There is strong evidence that obesity is associated with poor vitamin D status due to fat sequestration.[48] In fact, compared with the non-obese, higher doses of vitamin D are required in obese individuals. Also, regardless of dietary intake, weight loss results in increased serum $25(OH)D_3$ levels in overweight or obese women. However, some recent data suggest that low vitamin D predicts an accelerated increase in fat mass and incident obesity.

The importance of vitamin D for muscle has been discussed at length. Interestingly, VDR expression in human muscle tissue decreases with aging, and vitamin D level is strongly correlated with muscle quality and lower muscle fat infiltration. Consistently, in the Longitudinal Aging Study Amsterdam, low vitamin D and high parathyroid hormone were strong, independent risk factors for loss of muscle strength and muscle mass (sarcopenia).[49] Low vitamin D levels also appear to be correlated with collagen deposition and are related to arterial stiffness.[50] The role of vitamin D in energetics has been less thoroughly investigated. A recent study demonstrated that vitamin D level correlates with maximal oxygen consumption, and that supplementation of vitamin D increases fitness in athletes. Interestingly, a recent study found that VDRs are highly concentrated in the cristae of the mitochondria inner membrane.[51] The role of vitamin D in homeostatic signaling is complex and pervasive. Vitamin D affects several pathways related to inflammation and cell proliferation, including inhibition of aromatase (conversion from testosterone to estrogens), COX2, prostaglandin receptor, NF-κB, HIF-1 and VEGF, and stimulation of IGFBP-3 and E-cadherin.[52] In addition, poor vitamin D status is a risk factor for several autoimmune diseases, confirming the role of this vitamin in immune function.

The role of vitamin D in neurologic function and pathology has been studied from multiple perspectives. Vitamin D receptors are found in high concentration in various areas of the brain. Low vitamin D levels have been associated with accelerated cognitive decline in epidemiological studies, and low vitamin D dietary intake is associated with increased risk of Alzheimer's disease and depression.[53,54] Mechanisms suggested for this association include β-amyloid formation/aggregation, dysregulation of GABAergic activity and NMDA receptor, and increased calcium entry into neurons. Overall, vitamin D appears to be implicated in many of the physiological and pathological changes that occur with aging. Whether vitamin D supplementation may positively impact the aging process remains unknown and will require further long-term intervention studies.

Vitamin D dietary reference intakes
Daniel Bikle (University of California and VA Medical Center, San Francisco) addressed the recent Institute of Medicine (IOM) recommendations for dietary intake of vitamin D and evaluated whether they were sufficient in the context of emerging data about the roles of vitamin D and the prevalence of deficiency. Recommendations for the appropriate amount of vitamin D to be taken for optimal health—be it for musculoskeletal strength or for the numerous nonskeletal presumptive benefits of vitamin D—remain contentious. There is general consensus that the earlier recommendation of 400 IU of vitamin D/day was insufficient. Recently an IOM expert panel established to formulate new guidelines recommended that 600 IU/day would suffice for the general public aged 1–70 years, with the dose increased to 800 IU/day for those older than 70 years. However, doses up to 4000 IU/day were considered safe.[55] The task force compiling the Endocrine

Society Clinical Practice Guidelines (ESCPG) reached similar conclusions.[56] Vitamin D status is best assessed by the circulating levels of 25(OH)D$_3$. There is considerable variation among individuals with regard to the serum level of 25(OH)D$_3$ achieved with a given dose of oral vitamin D, and it is the serum level that counts with respect to biologic effect. The IOM expert panel concluded that most Americans (97.5%) had adequate levels of vitamin D, based on their conclusion that a 25(OH)D$_3$ of 20 ng/mL or greater was sufficient, raising the question of whether supplementation was even needed.

This conclusion was based on data from the 2001–2004 NHANES survey[57] that primarily sampled Caucasians, with limited sampling in Northern latitudes during the winter, the time and region that people would most likely have lower vitamin D levels. In that survey the average 25(OH)D$_3$ level was over 20 ng/mL. People of color, especially African-Americans, have substantially lower levels of 25(OH)D$_3$. Moreover, a number of studies have shown that prevailing levels of 25(OH)D$_3$, especially in older subjects, the institutionalized, and in most other countries, are well below those documented in the NHANES survey.[58] As parts of the aging process, the skin becomes less able to make vitamin D for a given amount of UVB exposure, the kidneys become less able to produce the active metabolite 1,25(OH$_2$)D$_3$, and the intestine becomes less able to respond to the 1,25(OH$_2$)D$_3$ with respect to calcium absorption. Furthermore, the conclusion by the IOM expert panel, that 20 ng/mL is sufficient, differs from that reached by the task force compiling the ESCPG. This task force recommended a higher target level, namely 30 ng/mL (75 mM), and pointed out that many individuals with a variety of conditions, including obesity, dark skin, malabsorption, renal failure, lactose intolerance, ingestion of certain drugs, and the institutionalized, are likely not to achieve adequate 25(OH)D levels by ingesting only 600 IU/day. If 30 ng/mL is deemed the optimal level (and neither set of recommendations considers this level unsafe), then the percent of Americans that maintain the optimal level falls well below 50%.

A further complication in trying to achieve a given level of 25(OH)D$_3$ is that 25(OH)D$_3$ assays have not proven consistent, making decisions regarding supplementation all the more difficult. This situation is improving with the availability of standards from the National Institute of Standards and Technol-

ogy. However, those assays relying on binding proteins or antibodies tend to report lower 25(OH)D levels than assays employing chromatography/mass spectroscopy.[59] Moreover, the latter can separately determine 25(OH)D$_2$ and 25(OH)D$_3$ levels, as well as the level of 3-epimer of 25(OH)D$_3$, capabilities that are lacking in the binding protein/antibody assays. While the clinical significance of these differences in assays is not clear, they do complicate decision making when a fixed target level of 25(OH)D$_3$ is being sought. In practical terms, young, healthy, light-skinned individuals with good diets that include dairy products and who have adequate exposure to sunlight are not likely to be vitamin D deficient or need to be tested. Individuals of color, the aged, those with gastrointestinal disorders, the institutionalized, and those with poor nutrition should be tested and treated appropriately to achieve a 25(OH)D$_3$ level of at least 20 ng/mL, or perhaps closer to 30 ng/mL. Daily doses appear to be safer and more efficacious than large doses administered once or twice a year.

Conclusions

The conference "Vitamin D: Beyond Bone" explored a considerable range of functions for vitamin D and potential therapeutic implications outside of calcium homeostasis and bone health, highlighting the increased understanding of vitamin D, as well as how much is yet to be learned. The scope of vitamin D research and its therapeutic possibilities, as presented by researchers across many different disciplines, generated enthusiasm tempered with some skepticism. Continuing research is needed to better understand the relationships between vitamin D and extraskeletal health and to determine the optimal dose of vitamin D for individuals based on age, health conditions, and other factors.

Acknowledgment

The conference "Vitamin D: Beyond Bone," presented by the Sackler Institute for Nutrition Science at the New York Academy of Sciences, was sponsored by an unrestricted educational grant from Abbott Nutrition Health Institute.

Conflicts of interest

The authors declare no conflicts of interest.

References

1. Christakos, S. & H.F. DeLuca. 2011. Minireview: Vitamin D: is there a role in extraskeletal health? *Endocrinology* **152:** 2930–2936.

2. Dhawan, P., R. Wieder & S. Christakos. 2009. CCATT enhancer binding protein alpha is a molecular target of 1,25-dihydroxyvitamin D3 in MCF-7 breast cancer cells. *J. Biol. Chem.* **284:** 3086–3095.

3. Joshi, S., L.-C. Pantalena, X. K. Liu, *et al.* 2011. 1,25-Dihydroxyvitamin D3 ameliorates Th17 autoimmunity via transcriptional modulation of interleukin-17A. *Mol. Cell. Biol.* **31:** 3653–3669.

4. Liu, P.T. *et al.* 2006. Toll-like receptor triggering of a vitamin D-mediated human antimicrobial response. *Science* **311:** 1770–1773.

5. Hewison, M. 2011. Antibacterial effects of vitamin D. *Nat. Rev. Endocrinol.* **7:** 337–345.

6. Hewison, M. *et al.* 2003. Differential regulation of vitamin D receptor and its ligand in human monocyte-derived dendritic cells. *J. Immunol.* **170:** 5382–5390.

7. Jeffery, L.E. *et al.* 2012. Availability of 25-Hydroxyvitamin D3 to APCs Controls the Balance between Regulatory and Inflammatory T Cell Responses. *J. Immunol.* **189:** 5155–5164.

8. Gardner, D.G., S. Chen, D.J. Glenn & W. Ni. 2011. Vitamin D and the Cardiovascular System. In: *Vitamin D*, 3rd Edition. D.J.W. Feldman, Pike and J.S. Adams, eds: 541–563. Elsevier: New York.

9. Li, Y.C., J. Kong, M. Wei, *et al.* 2002. 1,25-dihydroxyvitamin D(3) is a Negative Endocrine Regulator of the Renin-Angiotensin System. *J. Clin. Invest.* **110:** 229–238.

10. Chen, S. & D.G. Gardner. Liganded Vitamin D Receptor Displays Anti-Hypertrophic Activity in the Murine Heart. *J. Steroid. Biochem. Mol. Biol.*, in press.

11. Chen, S., C.S. Law, C.L. Grigsby, *et al.* 2011. Cardiomyocyte-specific deletion of the vitamin D receptor gene results in cardiac hypertrophy. *Circulation* **124:** 1838–1847.

12. Hollis, B.W., D. Johnson, T.C. Hulsey, *et al.* 2011. Vitamin D supplementation during pregnancy: double-blind, randomized clinical trial of safety and effectiveness. *J. Bone. Miner.* **26**(10): 2341–2357. PMCID: 3183324.

13. Mahomed, K. & A.M. Gulmezoglu. 1999. *Vitamin D supplementation in pregnancy (Cochrane Review). Cochrane Database of Systematic Reviews.* John Wiley & Sons, Ltd. Chicester, UK.

14. De-Regil, L.M., C. Palacios, A. Ansary, *et al.* 2012. Vitamin D supplementation for women during pregnancy. *Cochrane Database of Systematic Reviews* **2:** CD008873.

15. Hollis, B.W. & C.L. Wagner. 2012. Vitamin D and pregnancy: skeletal effects, nonskeletal effects, and birth outcomes. *Calcif. Tissue. Int.*

16. Wagner, C.L., R. McNeil, S.A. Hamilton, *et al.* 2012. A randomized trial of vitamin D supplementation in 2 community health center networks in South Carolina. *Am. J. Obstet. Gynecol.*

17. Wagner, C.L., T.C. Hulsey, D. Fanning, *et al.* 2006. High-dose vitamin D3 supplementation in a cohort of breastfeeding mothers and their infants: a 6-month follow-up pilot study. *Breastfeed Med.* **1:** 59–70.

18. Wagner, C.L., C. Howard, T.C. Hulsey, *et al.* 2012. Preliminary Results of a Randomized Controlled Trial of Maternal Supplementation with 6400 IU Vitamin D/day Compared with Maternal & Infant Supplementation of 400 IU/day in Achieving Sufficiency in the Breastfeeding Mother-Infant Dyad. *Intl. Society Res. Human Milk Lactation.* in press (abstract).

19. Sergeev, I.N. 2009. Novel mediators of vitamin D signaling in cancer and obesity. *Immun. Endoc. Metab. Agents Med. Chem.* **9:** 153–158.

20. Song, Q. & I.N. Sergeev. 2012. Calcium and vitamin D in obesity. *Nutr. Res. Rev.* **25:** 130–141.

21. Sergeev, I.N. 2012. Vitamin D and cellular Ca2+ signaling in breast cancer. *Anticancer Res.* **32:** 299–302.

22. Sergeev, I.N. 2005. Calcium signaling in cancer and vitamin D. *J. Steroid Biochem. Mol. Biol.* **97:** 145–151.

23. Sergeev, I.N. 2009. 1,25-Dihydroxyvitamin D3 induces Ca^{2+}-mediated apoptosis in adipocytes via activation of calpain and caspase-12. *Biochem. Biophys. Res. Commun.* **384:** 18–24.

24. Janssens, W. *et al.* 2009. Vitamin D Beyond Bones in COPD: Time to Act. *Am. J. Respir Crit. Care Med.*

25. Janssens, W. *et al.* Vitamin D deficiency is highly prevalent in COPD and correlates with variants in the vitamin D-binding gene. *Thorax.* **65**(3): 215–220.

26. Romme, E.A. *et al.* 2012. Vitamin D status is associated with bone mineral density and functional exercise capacity in patients with chronic obstructive pulmonary disease. *Ann. Med.*

27. Persson, L.J. *et al.* 2012. Chronic obstructive pulmonary disease is associated with low levels of vitamin d. *PLoS One* **7**(6): e38934.

28. Forli, L. *et al.* 2004. Vitamin D deficiency, bone mineral density and weight in patients with advanced pulmonary disease. *J. Intern. Med.* **256**(1): 56–62.

29. Hopkinson, N.S. *et al.* 2008. Vitamin D receptor genotypes influence quadriceps strength in chronic obstructive pulmonary disease. *Am. J. Clin. Nutr.* **87**(2): 385–390.

30. Lehouck, A. *et al.* 2012. High doses of vitamin D to reduce exacerbations in chronic obstructive pulmonary disease: a randomized trial. *Ann. Intern. Med.* **156**(2): 105–114.

31. Black, P.N. & R. Scragg. 2005. Relationship between serum 25-hydroxyvitamin d and pulmonary function in the third national health and nutrition examination survey. *Chest* **128**(6): 3792–3798.

32. Kumar, V.A., D.A. Kujubu, J.J. Sim, *et al.* 2011. Vitamin D supplementation and recombinant human erythropoietin utilization in vitamin D-deficient hemodialysis patients. *J. Nephrol.* **24**(1): 98–105.

33. Andrews, N.C. 2004. Anemia of inflammation:the cytokine-hepcidin link. *J. Clin. Invest.* **113**(9): 1251–1253.

34. Howick, J., P. Glasziou & J.K. Aronson. 2009. The evolution of evidence hierarchies: what can Bradford Hill's 'guidelines for causation' contribute? *J. Royal Soc. Med.* **102:** 186–194.

35. Mitri, J., M.D. Muraru & A.G. Pittas. 2011. Vitamin D and type 2 diabetes: a systematic review. *Eur. J. Clin. Nutr.* **65:** 1005–1015.

36. Song, Y., L. Wang, A.G. Pittas, *et al.* Blood 25-hydroxyvitamin D concentration and incident Type 2

Diabetes: A meta-analysis of prospective studies diabetes care (in press).

37. Pittas, A.G. *et al.* 2007. The effects of calcium and vitamin D supplementation on blood glucose and markers of inflammation in nondiabetic adults. *Diabetes Care* **30:** 980–986.

38. von Hurst, P.R. *et al.* 2008. Study protocol–metabolic syndrome, vitamin D and bone status in South Asian women living in Auckland, New Zealand: a randomised, placebo-controlled, double-blind vitamin D intervention. *BMC Public Health* **8:** 267.

39. Mitri, J. *et al.* 2011. Effects of vitamin D and calcium supplementation on pancreatic beta cell function, insulin sensitivity, and glycemia in adults at high risk of diabetes: the calcium and vitamin D for diabetes mellitus (CaDDM) randomized controlled trial. *Am. J. Clin. Nutr.* **94:** 486–494.

40. Boland, R. 1986. Role of vitamin D in skeletal muscle function. *Endocr. Rev.* **7:** 434–448.

41. Gallagher, J.C. 2004. The effects of calcitriol on falls and fractures and physical performance tests. *J. Steroid Biochem. Mol. Biol.* **89–90:** 497–501.

42. Bischoff, H.A., M. Borchers, F. Gudat, *et al.* 2001. In situ detection of 1,25-dihydroxyvitamin D3 receptor in human skeletal muscle tissue. *Histochem. J.* **33:** 19–24.

43. Boland, R. 2005. Vitamin D and Muscle. In: *Vitamin D.* 2nd. Edition; D. Feldman, G. Glorieux & W. Pike, eds: 885–898. Academic Press. CA, USA. Chapter 55.

44. Boland, R. 2011. VDR activation of intracellular signaling pathways in skeletal muscle. *Mol. Cell. Endocrinol.* **34:** 11–16.

45. Boland, R., A.R. de Boland, C. Buitrago, *et al.* 2002. Non-genomic stimulation of tyrosine phosphorylation cascades by 1,25(OH)2D3 by VDR-dependent and -independent mechanisms in muscle cells. *Steroids* **67:** 477–482.

46. Capiati, D., S. Benassati & R. Boland. 2002. 1,25(OH)2-vitamin D3 induces translocation of the vitamin D receptor (VDR) to the plasma membrane in skeletal muscle cells. *J. Cell. Biochem.* **86:** 128–135.

47. Houston, D.K., R.H. Neiberg, J.A. Tooze, *et al.* 2013. Low 25-hydroxyvitamin D predicts the onset of mobility limitation and disability in community-dwelling older adults: The health ABC study. *J. Gerontol. A Biol. Sci. Med. Sci.* **68**(2): 181–187.

48. Rock, C.L., J.A. Emond, S.W. Flatt, *et al.* 2012. Weight loss is associated with increased serum 25-hydroxyvitamin D in overweight or obese women. *Obesity.* **20**(11): 2296–2301.

49. Visser, M., D.J. Deeg, P. Lips, Longitudinal Aging Study Amsterdam. 2003. Low vitamin D and high parathyroid hormone levels as determinants of loss of muscle strength and muscle mass (sarcopenia): the longitudinal aging study amsterdam. *J. Clin. Endocrinol. Metab.* **88**(12): 5766–5772.

50. Giallauria, F., Y. Milaneschi, T. Tanaka, *et al.* 2012. Arterial stiffness and vitamin D levels: the Baltimore longitudinal study of aging. *J. Clin. Endocrinol. Metab.* **97**(10): 3717–3723.

51. Silvagno, F., E. De Vivo, A. Attanasio, *et al.* 2010. Mitochondrial localization of vitamin D receptor in human platelets and differentiated megakaryocytes. *PLoS One.* **5**(1): e8670.

52. Krishnan, A.V. & D. Feldman. 2011. Mechanisms of the anticancer and anti-inflammatory actions of vitamin D. *Annu. Rev. Pharmacol. Toxicol.* **51:** 311–336.

53. Llewellyn, D.J., I.A. Lang, K.M. Langa, *et al.* 2010. Vitamin D and risk of cognitive decline in elderly persons. *Arch. Intern. Med.* **170**(13): 1135–1141.

54. Milaneschi, Y., M. Shardell, A.M. Corsi, *et al.* 2010. Serum 25-hydroxyvitamin D and depressive symptoms in older women and men. *J. Clin. Endocrinol. Metab.* **95**(7): 3225–3233.

55. Institute of Medicine. 2011. *Dietary reference intakes for calcim and vitamin D.* Washington DC: The National Academies Press.

56. Holick, M.F., N.C. Binkley, H.A. Bischoff-Ferrari, *et al.* 2011. Evaluation, treatment and prevention of vitamin D deficiency: an endocrine society clinical practice guideline. *J. Clin. Endocrinol. Metab.* **96:** 1911–1930

57. Looker, A.C., C.L. Johnson, D.A. Lacher, *et al.* 2011. Vitamin D status: United States, 2001–2006. National Center for Health Statistics Data Brief no 59.

58. Kuchuk, N.O., N.M. van Schoor, S.M. Pluijm, *et al.* 2009. Vitamin D status, parathyroid function, bone turnover, and BMD in postmenopausal women with osteoporosis: global perspective. *J. Bone. Min. Res.* **24:** 693–701.

59. Binkley, N., D.C. Krueger, S. Morgan & D. Wiebe. 2010. Current status of clinical 25-hydroxyvitamin D measurement: an assessment of between-laboratory agreement. *Clin. Chim. Acta.* **411:** 1976–1982

60. Christakos, S., P. Dhawan, Y. Liu, *et al.* 2003. New Insights into the mechanisms of vitamin D action. *J. Cell. Biochem.* **88:** 695–705.